# Introduction to Astronomy and Cosmology

# Introduction to Astronomy and Cosmology

## Ian Morison
*University of Manchester, UK*

A John Wiley and Sons, Ltd., Publication

*Registered office*
John Wiley & Sons Ltd, The Atrium, Southern Gate, Chichester, West Sussex,
PO19 8SQ, United Kingdom

For details of our global editorial offices, for customer services and for information about how to apply for permission to reuse the copyright material in this book please see our website at www.wiley.com.

*Library of Congress Cataloging-in-Publication Data*

Morison, Ian, 1943–
    Introduction to astronomy and cosmology / Ian Morison.
        p.   cm.
    Includes bibliographical references and index.
    ISBN 978-0-470-03333-3 (cloth) — ISBN 978-0-470-03334-0 (pbk. : alk.paper)
    1. Astronomy—Textbooks.   2. Cosmology—Textbooks.   I. Title.
    QB43.3.M67 2008
    520—dc22

                                                        2008029112

A catalogue record for this book is available from the British Library.
Set in 8/10 pt Photina by Integra Software Services Pvt. Ltd. Pondicherry,
India.

*To the memory of my father, Archibald, who inspired my love of astronomy, to Bernard Lovell who made it possible for me to pursue that love and to my wife, Judy, with love.*

# Contents

## Chapter 4: Extra-solar Planets          135

## Chapter 5: Observing the Universe          153

## Chapter 6: The Properties of Stars    205

## Chapter 9: Cosmology – the Origin and Evolution of the Universe      301

# *Preface*

This textbook arose out of the lecture course that the author developed for first year physics and astronomy undergraduates at the University of Manchester. When it was proposed that all the students should undertake the course, not just those who had come to study astrophysics, several of the physics staff felt that it would not be appropriate for the physics students. But this view was countered with the fact that astronomy is a wonderful showcase for physics and this text covers aspects of physics ranging through Newton's and Einstein's theories of gravity, particle and nuclear physics and even quantum mechanics.

Not all of the material covered by the course was examinable; in particular the descriptions of the planets in our solar system and the background to some of the key discoveries of the last century. However, the author believes that this helps to give life to the subject and so these parts of the course have not been left out. Wherever possible, calculations have been included to illustrate all aspects of the book's material, but the level of mathematics required is not high and should be well within the capabilities of first year undergraduates. The questions with each chapter have come from course examination papers and tutorial exercises and should thus be representative of the type of questions that might be asked of students studying an astronomy course based on this book.

Some textbooks are rightly described as "worthy but dull". It is the author's earnest hope that this book would not fit this description and that, as well a conveying the basics of astronomy in an accessible way, it will be enjoyable to read. If, perhaps, the book could inspire some who have used it to continue their study of astronomy so that, in time, they might themselves contribute to our understanding of the universe, then it would have achieved all that its author could possibly hope for.

## Acknowledgements

The author would like to thank those who have helped this book to become a reality: the students who tested the questions and commented on the course material, my colleagues Phillipa Browning, Neil Jackson, Michael Peel and Peter Millington who carefully read through drafts of the text and the team at Wiley; Zoe Mills, Gemma Valler, Wendy Harvey, Andy Slade and Richard Davies who have provided help and encouragement during its writing and production.

No matter how hard we have all tried, the text may well contain some mistakes – for which the author takes full responsibility! To help eradicate them, should there be future editions, he would be most grateful if you could send comments and corrections via the website: http://www.jb.man.ac.uk/public/im/astronomy.html

This will provide additional supporting material for the book and a (hopefully small) list of corrections.

# Author's Biography

Ian Morison began his love of astronomy when, at the age of 12, he made a telescope out of lenses given to him by his optician. He went on to study Physics, Mathematics and Astronomy at Oxford and in 1970 was appointed to the staff of the University of Manchester where he now teaches astronomy, computing and electronics.

He is a past president of the Society for Popular Astronomy, one of the UK's largest astronomical societies. He remains on the society's council and holds the post of instrument advisor helping members with their choice and use of Telescopes.

He lectures widely on astronomy, has co-authored books for amateur astronomers and writes regularly for the two UK astronomy magazines. He also writes a monthly sky guide for the Jodrell Bank Observatory's web site and produces an audio version as part of the Jodrell Bank Podcast. He has contributed to many television programmes and is a regular astronomy commentator on local and national radio. Another activity he greatly enjoys is to take amateur astronomers on observing trips such as those to Lapland to see the Aurora Borealis and on expeditions to Turkey and China to observe total eclipses of the Sun.

In 2003 the Minor Planets Committee of the International Astronomical Union named asteroid 15,727 in his honour, citing his work with MERLIN, the world's largest linked array of radio telescopes, and that in searching for intelligent life beyond our Solar System in Project Phoenix. In 2007 he was appointed to the post of Gresham Professor of Astronomy. Dating from 1597, this is the oldest astronomy professorship in the world and was once held by Christopher Wren.

## Chapter 1

# Astronomy, an Observational Science

## 1.1    Introduction

Astronomy is probably the oldest of all the sciences. It differs from virtually all other science disciplines in that it is not possible to carry out experimental tests in the laboratory. Instead, the astronomer can only observe what he sees in the Universe and see if his observations fit the theories that have been put forward. Astronomers do, however, have one great advantage: in the Universe, there exist extreme states of matter which would be impossible to create here on Earth. This allows astronomers to make tests of key theories, such as Albert Einstein's General Theory of Relativity. In this first chapter, we will see how two precise sets of observations, made with very simple instruments in the sixteenth century, were able to lead to a significant understanding of our Solar System. In turn, these helped in the formulation of Newton's Theory of Gravity and subsequently Einstein's General Theory of Relativity – a theory of gravity which underpins the whole of modern cosmology. In order that these observations may be understood, some of the basics of observational astronomy are also discussed.

## 1.2    Galileo Galilei's proof of the Copernican theory of the solar system

One of the first triumphs of observational astronomy was Galileo's series of observations of Venus which showed that the Sun, not the Earth, was at the centre of the Solar System so proving that the Copernican, rather than the Ptolemaic, model was correct (Figure 1.1).

In the Ptolemaic model of the Solar System (which is more subtle than is often acknowledged), the planets move around circular 'epicycles' whose centres move around the Earth in larger circles, called deferents, as shown in Figure 1.2. This enables it to account for the 'retrograde' motion of planets like Mars and Jupiter when they appear to move backwards in the sky. It also models the motion of Mercury and Venus. In their case, the deferents, and hence the centre of their

*Introduction to Astronomy and Cosmology*   Ian Morison
© 2008 John Wiley & Sons, Ltd

**Figure 1.1** Galileo Galilei: a portrait by Guisto Sustermans. Image: Wikipeda Commons.

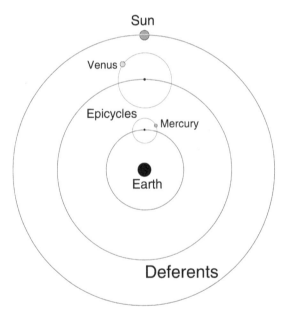

**Figure 1.2** The centre points of the epicycles for Mercury and Venus move round the Earth with the same angular speed as the Sun.

epicycles, move around the Earth at the same rate as the Sun. The two planets thus move around in circular orbits, whose centres lie on the line joining the Earth and the Sun, being seen either before dawn or after sunset. Note that, as Mercury stays closer to the Sun than Venus, its deferent and epicycle are closer than that of Venus – in the Ptolemaic model, Mercury is the closest planet to the Earth!

In the Ptolemaic model, Venus lies between the Earth and the Sun and hence it must always be lit from behind, so could only show crescent phases whilst its angular size would not alter greatly. In contrast, in the Copernican model Venus orbits the Sun. When on the nearside of the Sun, it would show crescent phases whilst, when on its far side but still visible, it would show almost full phases. As its distance from us would change significantly, its angular size (the angle subtended by the planet as seen from the Earth) would likewise show a large change.

Figure 1.3 shows a set of drawings of Venus made by Galileo with his simple refracting telescope. They are shown in parallel with a set of modern photographs which illustrate not only that Galileo showed the phases, but that he also drew the changing angular size correctly. These drawings showed precisely what the Copernican model predicts: almost full phases when Venus is on the far side of the Sun and a small angular size coupled with thin crescent phases, having a significantly larger angular size, when it is closest to the Earth.

Galileo's observations, made with the simplest possible astronomical instrument, were able to show which of the two competing models of the Solar System was correct. In just the same way, but using vastly more sophisticated instruments, astronomers have been able to choose between competing theories of the Universe – a story that will be told in Chapter 9.

**Figure 1.3** Galileo's drawings of Venus (top) compared with photographs taken from Earth (bottom).

## The celestial sphere and stellar magnitudes

Looking up at the heavens on a clear night, we can imagine that the stars are located on the inside of a sphere, called the celestial sphere, whose centre is the centre of the Earth.

### 1.3.1    *The constellations*

As an aid to remembering the stars in the night sky, the ancient astronomers grouped them into constellations; representing men and women such as Orion, the Hunter, and Cassiopeia, mother of Andromeda, animals and birds such as Taurus the Bull and Cygnus the Swan and inanimate objects such as Lyra, the Lyre. There is no real significance in these stellar groupings – stars are essentially seen in random locations in the sky – though some patterns of bright stars, such as the stars of the 'Plough' (or 'Big Dipper') in Ursa Major, the Great Bear, result from their birth together in a single cloud of dust and gas.

The chart in Figure 1.4 shows the brighter stars that make up the constellation of Ursa Major. The brightest stars in the constellation (linked by thicker lines) form what in the UK is called 'The Plough' and in the USA 'The Big Dipper', so called after the ladle used by farmers' wives to give soup to the farmhands at lunchtime. On star charts the brighter stars are delineated by using larger diameter circles

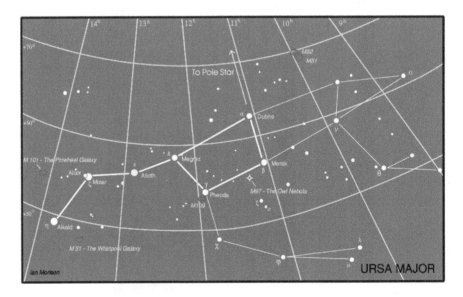

**Figure 1.4** The constellation of Ursa Major – the Great Bear.

which approximates to how stars appear on photographic images. The grid lines define the positions of the stars on the celestial sphere as will be described below.

### 1.3.2     *Stellar magnitudes*

The early astronomers recorded the positions of the stars on the celestial sphere and their observed brightness. The first known catalogue of stars was made by the Greek astronomer Hipparchos in about 130–160 BC. The stars in his catalogue were added to by Ptolomy and published in 150 AD in a famous work called the Almagest whose catalogue listed 1028 stars. Hipparchos had grouped the stars visible with the unaided eye into six magnitude groups with the brightest termed 1st magnitude and the faintest, 6th magnitude. When accurate measurements of stellar brightness were made in the nineteenth century it became apparent that, on average, the stars of a given magnitude were approximately 2.5 times brighter than those of the next fainter magnitude and that 1st magnitude stars were about 100 times brighter than the 6th magnitude stars. (The fact that each magnitude difference showed the same brightness ratio is indicative of the fact that the human eye has a logarithmic rather than linear response to light.)

In 1854, Norman Pogson at Oxford put the magnitude scale on a quantitative basis by defining a five magnitude difference (i.e., between 1st and 6th magnitudes) to be a brightness ratio of precisely 100. If we define the brightness ratio of one magnitude difference as $R$, then a 5th magnitude star will be $R$ times brighter than a 6th magnitude star. It follows that a 4th magnitude star will be $R \times R$ times brighter than a 6th magnitude star and a 1st magnitude star will be $R \times R \times R \times R \times R$ brighter than a 6th magnitude star. However, by Pogson's definition, this must equal 100 so $R$ must be the 5th root of 100 which is 2.512.

**The brightness ratio between two stars whose apparent magnitude differs by one magnitude is 2.512.**

Having defined the scale, it was necessary to give it a reference point. He initially used Polaris as the reference star, but this was later found to be a variable star and so Vega became the reference point with its magnitude defined to be zero. (Today, a more complex method is used to define the reference point.)

### 1.3.3     *Apparent magnitudes*

It should be noted that the observed magnitude of a star tells us nothing about its intrinsic brightness. A star that appears bright in the sky could either be a faint star that happens to be very close to our Sun or a far brighter star at a greater distance. As a result, these magnitudes are termed apparent magnitudes. The nominal apparent magnitudes relate to the brightness as observed with instruments having the same wavelength response as the human eye. As we shall see in Chapter 6, one can also measure the apparent magnitudes as observed in specific wavebands, such as red or blue, and such measurements can tell us about the colour of a star.

Some stars and other celestial bodies, such as the Sun, Moon and planets are much brighter than Vega and so can have negative apparent magnitudes. Magnitudes can also have fractional parts as, for example, Sirius which has a magnitude of $-1.5$. Figure 1.5 gives the apparent magnitudes of a range of celestial bodies from the brightest, the Sun, to the faint dwarf planet, Pluto.

### 1.3.4    *Magnitude calculations*

From the logarithmic definition of the magnitude scale two formulae arise.

The first gives the brightness ratio, $R$, of two objects whose apparent magnitude differs by a known value $\Delta m$:

$$R = 2.512^{\Delta m} \qquad (1.1)$$

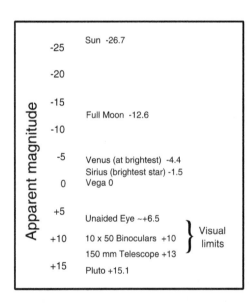

**Figure 1.5  Some examples of apparent magnitudes.**

The second gives the magnitude difference between two objects whose brightness ratio is known. We can derive this from the first as follows:

Taking logarithms to the base 10 of both sides of Equation (1.1) gives:

$$
\begin{aligned}
\mathrm{Log}_{10}\, R &= \mathrm{Log}_{10}(2.512) \times \Delta m \\
\mathrm{Log}_{10}\, R &= 0.4 \times \Delta m \\
\Delta m &= \mathrm{Log}_{10}\, R/0.4 \\
\Delta m &= 2.5 \times \log_{10} R
\end{aligned}
$$

As an example, using values from Figure 1.5, let us calculate how much brighter the Sun is than the Moon.

The difference in magnitudes is $26.7 - 12.6 = 14.1$, so

$$
\begin{aligned}
R &= 2.512^{14.1} \\
&= 436\,800
\end{aligned}
$$

The Sun is ~440 000 times brighter than the full Moon.

This perhaps emphasizes the fact that the eye can cope with an incredibly wide range in brightness: we can see a surprising amount with the light of the full Moon and yet can cope with the light on a bright sunny beach.

Consider a second example: a star has a brightness which is 10 000 times less than Vega (magnitude 0). What is the magnitude of the star?

There is a quick way to do this: 10 000 is $100 \times 100$. However, a ratio of 100 in brightness is 5 magnitudes so this star must be 10 magnitudes fainter than Vega and will thus be 10th magnitude.

Using the formula:
$$
\begin{aligned}
\Delta m &= 2.5 \times \log_{10}(10\,000) \\
&= 2.5 \times 4 \\
&= 10
\end{aligned}
$$

gives the same result.

## The celestial coordinate system

The early star catalogues located the positions of the stars on the celestial sphere using a slightly different coordinate system than we do now. The modern coordinate system is analogous to the way in which we define positions on the surface of the Earth and uses the orientation of the Earth in space as its basis. The Earth's rotation axis is extended up and down to the points where it reaches our imaginary celestial sphere. The point where the axis meets the sphere directly above the North Pole is called the North Celestial Pole and that below the South Pole is the South Celestial Pole. If the Earth's equator is extended outwards it will cut the celestial sphere into two – into the northern and southern hemispheres – forming the Celestial Equator (see Figure 1.6).

There is one path around the celestial sphere that is of great importance: that of our Sun. If the Earth's rotation axis was at right angles to the plane of its orbit around the Sun, the Sun's path would trace out the Celestial Equator but, as the axis of the Earth's rotation is inclined to its orbital plane by an angle of 23.5°, the path of the Sun is a great circle, called the ecliptic, which is inclined by 23.5° to the Celestial Equator. The Sun spends half the year in the southern half of the celestial sphere and the other half in the northern. Its path thus crosses the Celestial Equator twice every year: once at the vernal equinox, on March 20 or 21, as it comes into the northern hemisphere and 6 months later when, at the autumnal equinox on September 22 or 23, it returns to the southern hemisphere.

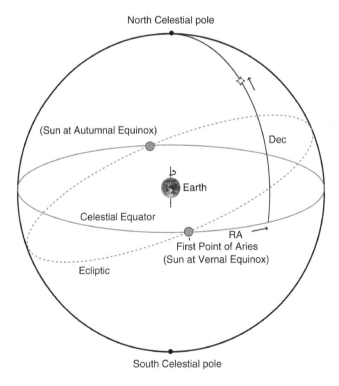

**Figure 1.6** The celestial sphere.

Just as a location on the Earth's surface has a 'latitude', defined as its angular distance from the equator towards the poles, so a star has a 'declination'" (Dec) given as an angle which is either positive (in the northern hemisphere) or negative (in the southern hemisphere). The 'Pole Star' in the northern sky is close to the North Celestial Pole at close to +90 declination and the region at the South Celestial Pole (where there is no bright star) is at −90 declination.

The second coordinate proves to be rather more difficult. On the Earth we define the position of a location round the Earth by its longitude. However, there has to be some arbitrary zero of longitude. It was sensible that the zero of longitude, called the Prime Meridian, should pass through a major observatory and that honour finally fell to the Royal Greenwich Observatory in London.

As referred to above, the path of the Sun gives two defined points along the Celestial Equator that might sensibly be used as the zero of Right Ascension (RA) – the points where the ecliptic crosses the Celestial Equator at the vernal and autumnal equinoxes. The point where the Sun moves into the northern hemisphere was chosen and was given the name 'The first point of Aries' as this was the constellation in which it lay. Star positions are measured eastwards around the celestial sphere from the first point in Aries to give the star's RA.

However, for reasons that will become apparent when we describe how star positions are measured, RA is not measured in degrees but in time, with 24 h equivalent to 360°. Hence, the celestial sphere is split into 24 segments each of 1 h and equivalent to 15° around the Celestial Equator.

**Angular measure**
A great circle measures 360° in angular extent.
Each degree is divided into 60 arcmin.
Each arcminute is divided into 60 arcsec.
There are then 3600 arcsec in 1°.
(Arcseconds and arcminutes can also be written as seconds of arc and minutes of arc, respectively.)

**1.5**

## Precession

Should you locate the point where the Sun crosses the ecliptic at the vernal equinox on a star chart (with position: RA = 0:00 h, Dec = 0.0°), you might be surprised to find that it is not in Aries, but in the adjacent constellation Pisces. This is the result of the precession of the Earth's rotation axis in just the same way that the axis of rotation of a spinning top or gyroscope is seen to precess. The precession rate is slow; one rotation every ~26 000 years, but its effect over the centuries is to change the positions of stars as measured with the co-ordinate system described above, which is fixed to the Earth. Consequently, a star chart is only valid for one specific date. Current star charts show the positions of stars as they were at the start of the millennium and will state 'Epoch 2000' in their titles. One result of precession is that the Pole Star is only close to the North Celestial Pole at this particular moment in time in the precession cycle (Figure 1.7). In ~12 000 years, the bright star Vega will be near the North Celestial Pole instead (though by no means as close). It also means that constellations currently not observable from the UK will become visible above the southern horizon.

Interestingly, it is stars in the part of the sky that was visible to ancient astronomers and which were thus included in the constellations that enable us to estimate not only the time but also the latitude from which the constellations were delineated and named.

A region of about 36° radius in the southern sky did not contain any of the original 48 constellations implying that this region was invisible to those who

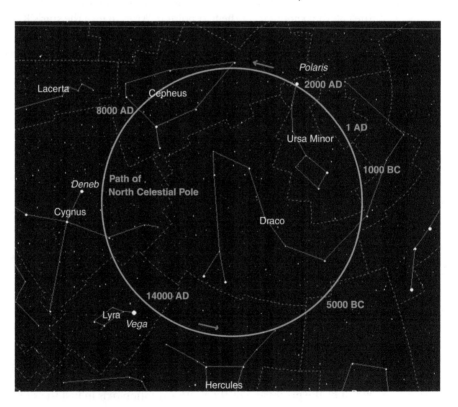

**Figure 1.7** The path of the North Celestial Pole through the heavens.

mapped the sky. This is precisely the region that would have been invisible to those living at a latitude of 36° north. Due to precession, the stars that would be hidden from view in this region will vary with time, and this enables us to give a date, about 2600–2900 BC, when the constellations were delineated. This origin would also explain the reason behind the long, thin constellation Hydra, the sea serpent, that now arcs across 95° of the sky – Hydra then followed the line of the Celestial Equator.

The ancient Greek poet, Aratus, described how some stars set as others rose into the night sky at opposite points on the horizon. Again, due to precession, such coincidences will also depend on the latitude of the observer and the time of observation. These observations also imply a latitude of about 36° north and a date about 2600 BC.

## 1.6     Time

Before one can appreciate the observations that lay behind a second observational triumph, which came about from the precise observations of the positions of the stars and planets by Tycho Brahe in the seventeenth century, it is necessary to understand how astronomers measure time.

### 1.6.1     *Local solar time*

For centuries, the time of day was directly linked to the Sun's passage across the sky, with 24 h being the time between one transit of the Sun across the meridian and that on the following day. This time standard is called 'Local Solar Time' and is the time indicated on a sundial. The time such clocks would show would thus vary across the UK, as noon is later in the west. It is surprising the difference this makes. In total, the UK stretches 9.55° in longitude from Lowestoft in the east to Mangor Beg in County Fermanagh, Northern Ireland in the west. As 15° is equivalent to 1 h, this is a time difference of just over 38 min!

### 1.6.2     *Greenwich mean time*

As the railways progressed across the UK, this difference became an embarrassment and so London or 'Greenwich' time was applied across the whole of the UK. A further problem had become apparent as clocks became more accurate: due to the fact that the Earth's orbit is elliptical the length of the day varies slightly. Thus, 24 h, as measured by clocks, was defined to be the *average* length of the day over 1 year. This time standard became known as Greenwich Mean Time (GMT).

### 1.6.3    *The equation of time*

The use of GMT has the consequences that, during the year, our clocks get in and out of step with the Sun. The difference between GMT and the Local Solar Time at Greenwich is called the 'Equation of Time'(Figure 1.8). The result is that the Sun is not always due south at noon, even in London, and the Sun can transit (cross the meridian) up to 16 min 33 s before noon as measured by a clock giving GMT and up to 14 min 6 s afterwards. This means that sunrise and sunset are not usually symmetrically centred on midday and this does give a noticeable effect around Christmas time. Though the shortest day is on December 21, the Winter Solstice, the earliest sunset is around December 10 and the latest sunrise does not occur until January 2, so the mornings continue to get darker for a couple of weeks after December 21 whilst, by the beginning of January, the evenings are appreciably longer.

### 1.6.4    *Universal time*

Greenwich Mean Time was formally replaced by Universal Time (UT) in 1928 (although the title has not yet come into common usage) but was essentially the same as GMT until 1967 when the definition of the second was changed! Prior to this, 1 s was defined as 1/86 400th of a mean day as determined by the rotation of the Earth. The rotation rate of the Earth was thus our fundamental time standard. The problem with this definition is that, due to the tidal forces of the Moon,

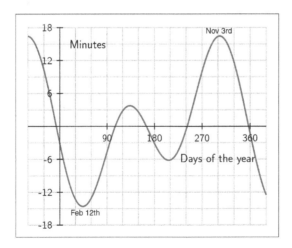

**Figure 1.8**  The 'Equation of Time' – the difference between GMT and Local Solar Time at Greenwich Observatory. Image: Wikipedia Commons.

the Earth's rotation rate is gradually slowing and, as a consequence, the length of time defined by the second was increasing. Hence, in 1967, a new definition of the second was made:

**The second is the duration of 9 192 631 770 periods of the radiation corresponding to the transition between the two hyperfine levels of the ground state of the caesium 133 atom.**

Thus our clocks are now related to an Atomic Time standard which uses caesium beam frequency standards to determine the length of the second.

This has not stopped the Earth's rotation from slowing down, and so very gradually the synchronization between the Sun's position in the sky and our clocks will be lost. To overcome this, when the difference between the time measured by the atomic clocks and the Sun (as determined by the Earth's rotation rate) differs by around a second, a leap second is inserted to bring solar and atomic time back in step. This is usually done at midnight on New Year's Eve or June 30. Since the time definition was changed, 22 leap seconds have had to be added, about one every 18 months, but there were none between 1998 and 2005 showing the slowdown is not particularly regular. Leap seconds are somewhat of a nuisance for systems such as the Global Positioning System (GPS) Network and there is pressure to do away with them which is, not surprisingly, opposed by astronomers! If no correction was made and the average slow down over the last 39 years of 0.56 of a second per year continues, then in 1000 years UT and solar time would have drifted apart by ~9 min.

## 1.6.5    *Sidereal time*

If one started an electronic stop watch running on UT as the star Rigel, in Orion, was seen to cross the meridian and stopped it the following night when it again crossed the meridian, it would be found to read 23 h, 56 min and 4.09 s, not 24 h. This period is called the sidereal day and is the length of the day as measured with respect to the apparent rotation of the stars.

Why does the sidereal day have this value? Imagine that the Earth was not rotating around its axis and we could observe from the dark side of the Earth facing away from the Sun. At some point in time we would see Rigel due south. As the Earth moves around the Sun, Rigel would be seen to move towards the west and, 3 months later, would set from view. Six months later after setting in the west, it would be seen to rise in the east and precisely 1 year later we would see it due south again. So, in the absence of the Earth's rotation, Rigel would appear to make one rotation of the Earth in 1 year and so the sidereal day would be 1 Earth year. In reality, during this time, the Earth has made ~365 rotations so, in relation to the star Rigel (or any other star), the Earth has made a total of ~365 + 1 rotations in 1 year and hence there are ~366 sidereal days in 1 year. The sidereal day is thus a little shorter and is approximately 365/366 of an Earth day.

The difference would be ~1/366 of a day or 1440/366 min giving 3.93 min or 3 min 55.8 s. The length of the sidereal day on this simplified calculation is thus approximately 23 h 56 min 4.2 s, very close to the actual value.

### 1.6.6    *An absolute time standard – cosmic time*

This section is not really necessary for the development of this chapter, but the ideas described here are very interesting and allow some discussion of Einstein's theories of relativity and the fact that, only recently, we have been able to consider how time measured by us on Earth relates to the timescale of the universe.

In 1905, Albert Einstein, then working in the Berne Patent Office, published his paper on the Special Theory of Relativity. Perhaps one of the most well known aspects of this theory is that moving clocks appear to run slow when compared with a clock at rest with an observer – a phenomenon called time dilation. This prediction has been proven by flying highly accurate atomic clocks around the world and has to be taken into account in the GPS system used for navigation. (See the box at the end of this section.)

As time is relative can we actually define a time standard with which to observe the evolution of the universe? One could, perhaps, define what might be called cosmic time as that measured by a clock that is stationary with respect to the universe as a whole. How would this time relate to clocks on Earth? We know that the Earth is moving around the Sun, and that the Sun is moving around the centre of our Milky Way Galaxy once every ~220 million years. Can we measure how fast the Solar System is moving with respect to the universe? Perhaps surprisingly, we can.

In Chapter 9 we will see how observations have been made of what is called the Cosmic Microwave Background (CMB) – radiation that originated near the time of its origin and which now pervades the whole universe. This radiation is very largely composed of a mix of long wavelength infrared and very short wavelength radio waves – it has a 'black body spectrum' that will be discussed in Chapter 2. For simplicity, just suppose that it is made up of only one wavelength and that the Solar System is moving in a certain direction with respect to this radiation. The Doppler effect will alter the apparent wavelength that we observe so that, when looking along the direction in which the Solar System is moving, it will be blue shifted and appear to have a shorter wavelength. Conversely, in the opposite direction, the radiation will appear to be red shifted and have a longer wavelength. From very precise measurements of the CMB we now know that we are moving towards the constellation Leo at a speed of ~650 km s$^{-1}$ (2 340 000 km h$^{-1}$ or about 0.22% of the speed of light). This is our speed with respect to the universe as a whole.

We can thus calculate how the time of a clock at rest with the universe – measuring cosmic time – will differ from our clocks. To do this we need to derive the formula that determines the observed time dilation as a function of relative speed. This is not difficult if we can imagine a very simple 'clock'.

The clock is made by reflecting a photon back and forth between a pair of perfect mirrors separated by a distance, $d$, as seen in Figure 1.9a. Our 'tick' happens every time the photon reflects off the lower mirror and so the photon will travel a distance $2d$ between each tick. Our fundamental time period, $t_1$, will thus be given by:

$$t_1 = 2d/c$$

Suppose we observe such a clock moving past us at speed $v$. We will see the situation in Figure 1.9b. As seen from our point of view, the photon will have to travel a longer distance, $l$, between each tick. This distance is given by:

$$l = [(2d)^2 + (vt_2)^2]^{1/2}$$

The time interval between each tick, $t_2$, will then be given by:

$$t_2 = l/c = [(4d^2 + v^2t_2^2)/c^2]^{1/2}$$

($c$ has been squared and put inside the square root.)

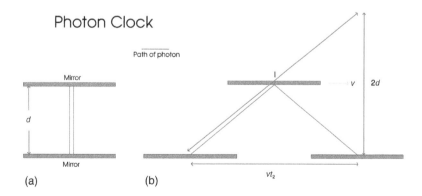

## Photon Clock

(a)  (b)

**Figure 1.9** (a) A photon clock at rest with the observer and (b) a photon clock moving at a speed $v$ with respect to the observer.

Squaring both sides and cross multiplying gives;

$$t_2{}^2c^2 = 4d^2 + v^2t_2{}^2$$

We can now relate $t_2$ and $t_1$ to $v$ by substituting for $d$ from above using $d^2 = t_1{}^2c^2/4$, giving:

$$t_2{}^2c^2 = t_1{}^2c^2 + v^2t_2{}^2$$

and

$$t_2{}^2(c^2 - v^2) = t_1{}^2c^2$$

so, finally,

$$t_2/t_1 = [c^2/(c^2 - v^2)]^{1/2}$$

or,

$$t_2/t_1 = 1/[1 - (v^2/c^2)]^{1/2}$$

This is the time dilation formula, giving the ratio of time intervals as a function of the relative speed $v$.

We can now enter our speed with respect to the universe, $650 \, \mathrm{km \, s^{-1}}$, into this equation and get the ratio 1.0000023. This is exceedingly small so, to a very good approximation, our clocks can be used to measure the timescale of the universe.

**Relativity and the Global Positioning System**

Travelling around the globe at a speed of $3.9 \, \mathrm{km \, s^{-1}}$, the atomic clocks providing the time signals in the GPS satellite constellation will lose ~$7.2 \, \mu\mathrm{s \, day^{-1}}$ as measured by clocks on the ground. (You could try this calculation.) It should however be pointed out that there is an even greater effect due to the fact that the GPS clocks are in a weaker gravitational field. This makes them run fast compared with clocks on the ground by $45.9 \, \mu\mathrm{s \, day^{-1}}$. Combining the two effects gives a net offset of $+38.7 \, \mu\mathrm{s \, day^{-1}}$. To account for this, the frequency standards on board the GPS satellites are given a rate offset prior to launch, making them run slightly slow – they are set to $10.22999999543 \, \mathrm{MHz}$ instead of $10.23 \, \mathrm{MHz}$.

**1.7**

## A second major observational triumph: the laws of planetary motion

We are now in a position to describe the observations which led to a further major improvement in our understanding of the Solar System.

### 1.7.1        *Tycho Brahe's observations of the heavens*

In 1572, Tycho Brahe, a young Danish nobleman whose passion was astronomy, observed a supernova (a very bright new star) in the constellation of Cassiopeia. His published observations of the new 'star' shattered the widely held belief that the heavens were immutable and he became a highly respected astronomer.

He realised that in order to show when further changes in the heavens might take place it was vital to have a first class catalogue of the visible stars. Four years later, Tycho was given the Island of Hven by the King of Denmark and money to build a castle that he called Uraniborg, named after Urania, the Greek Goddess of the heavens. In the castle's grounds, he built a semi-underground observatory called Stjerneborg. For a period of 20 years, his team of observers made positional measurements of the stars and, critically important, the planets.

Figure 1.10b shows the observatory and Figure 1.11 indicates how the measurements were made. An observer sighted a star (or planet) through a small window on a south facing wall. He did two things. First, he was able to indicate to his assistants when the star crossed the meridian. (The meridian is the half-circle that runs across the sky through the zenith between the north and south poles and intersects the horizon due south.) Secondly, by using a giant quadrant equipped with vernier scales, he was able to measure the elevation (angular height above the horizon) of the star at the moment of transit. One assistant is standing beside the clock at the lower right of the diagram to measure the time at which the star transits and a second assistant is seated at a table at the lower left who would then note the elevation of the star and time of transit in the logbook. From Figure 1.12 you can see that, given its observed elevation and the latitude of the observatory, the declination of the star can be found directly.

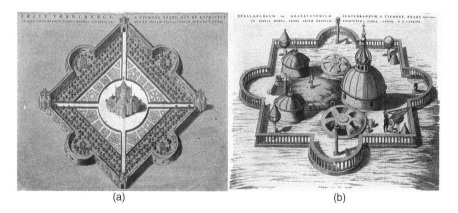

(a)                                                                 (b)

**Figure 1.10** Tycho Brahe's castle, Uraniborg (a), and his observatory, Stjerneborg (b), on the Island of Hven. Image: Wikipedia Commons.

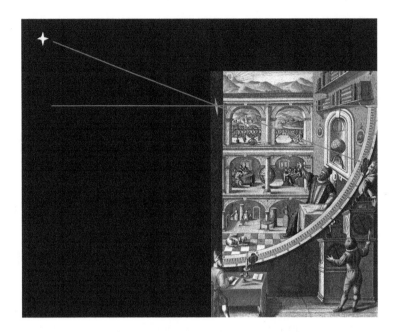

**Figure 1.11** Observing the elevation of a star as it transited due south. Observatory image: Wikipedia Commons. Note, on the original, the window is too high.

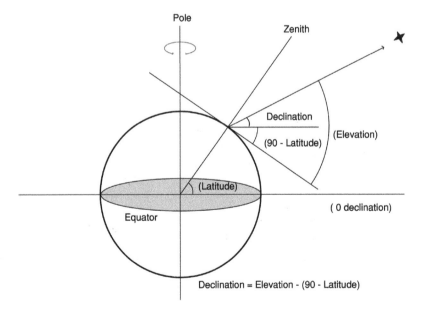

**Figure 1.12** The geometry showing how the declination of a star is derived. The zenith is the point directly above the observer.

The time of the transit gives the RA. Let us now suppose that Tycho Brahe set his clock to measure sidereal time. If he now set his clock to read 0:00 h at the time when the First Point of Aries crossed the meridian, then the time when a star crossed the meridian would directly give the RA! You can now see how the convention that RA is measured in units of time and increases to the east came about.

Today, transit telescopes, such as that at the Royal Greenwich Observatory, are used to measure stellar positions. These can only observe due south and, like Tycho's quadrant, are used to measure the elevation of a star as it transits the meridian. It is observations made by this type of telescope that have shown that, gradually, the rotation rate of the Earth has been slowing down, and they are used to decide when a 'leap second' should be added.

The time of transit would now be measured in UT but, given the value of the sidereal time at the previous midnight, 'GST at midnight' (GST is Greenwich Sidereal Time), the sidereal time at the moment of observation, and hence the RA can be easily calculated as follows.

Suppose a star is observed to transit at 02:23:36 UT and given that the GST at the previous midnight, as found in the Nautical Almanac, was 19:16:21, a ball-park figure of the RA could be found by just adding these two times together to give 21:39:57. However, this simple calculation neglects to account for the fact that sidereal seconds are shorter than UT seconds, so that the increase in sidereal time since midnight will be slightly greater than the increase in UT. To the nearest second, there are $(23 \times 3600) + (56 \times 60) + 4 = 86\,164$ UT seconds in 1 sidereal day. A sidereal second is thus $86\,164/86\,400$ times shorter than a UT second ($86\,400 = 24 \times 3600$).

> The accurate calculation:
>
> The transit was $7200 + 1380 + 36 = 8616$ UT seconds after midnight.
> This would equate to $8616 \times 86\,400/86\,164 = 8639$ sidereal seconds.
> This is 02:23:59 as measured in sidereal time.
> The RA of the star would thus be $19:16:21 + 02:23:59 = 21:40:20$.
>
> (This is 23 s greater in RA than the ball-park figure.)

Not only had Tycho produced a star catalogue 10 times more precise than any previous astronomer – the errors of the 777 star positions never exceeded 4 arcmin – he had also charted the movement of the planets during the 20 year period of his observations. It was these planetary observations that led to the second major triumph of observational astronomy in the sixteenth and seventeenth centuries: Kepler's three laws of planetary motion.

### 1.7.2    *Johannes Kepler joins Tycho Brahe*

When King Frederik II died in 1588, Tycho lost his patron. The final observation at Hven was made in 1596 before Tycho left Denmark. After a year travelling around Europe he was offered the post of Imperial Mathematician to Rudolf II, the Holy Roman Emperor, and was installed in the castle at Benatky. It was here that a young mathematician, Johannes Kepler, came to work with him. Tycho gave him the task of solving the orbit of the planet Mars. Kepler thought that it would take him a few months. In fact, it took him several years!

There was a fundamental problem. The observations of Mars had been made from the Earth, which was itself in orbit around the Sun. Unless one knew the precise orbit of the Earth, one could not find the parameters of the Martian orbit. In what has been described as a stroke of genius, Kepler realized that every 687 days (the orbital period of Mars) Mars would return to exactly the same location in the Solar System, so observations of the Earth from Mars, made on a set of dates separated by 687 days, could be used to find the precise orbit of the Earth. (As could, of course, observations made on those days of Mars from the Earth.) Having first solved for the Earth's orbit, Kepler was then able to deduce the orbital parameters of Mars.

### 1.7.3    *The laws of planetary motion*

From the invaluable database of planetary positions provided by Tycho, Kepler was able to draw up his three empirical laws of planetary motion. The word 'empirical' indicated that these laws were not based on any deeper theory, but accurately described the observed motion of the planets. The first two were published in 1609 and the third in 1618.

The first law states that:

**Planets move in elliptical orbits around the Sun, with the Sun positioned at one focus of the ellipse.**

Figure 1.13 shows a planet in an elliptical orbit around the Sun and defines some of the terms associated with the orbit.

The second law states that:

**The radius vector – that is, the imaginary line joining the centre of the planet to the centre of the Sun – sweeps out equal areas in equal times.**

Figure 1.14 shows Kepler's Second Law graphically.

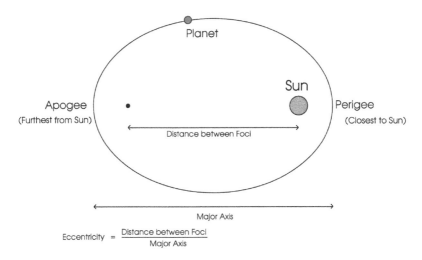

Apogee
(Furthest from Sun)

Perigee
(Closest to Sun)

Planet

Sun

Distance between Foci

Major Axis

Eccentricity  =  $\dfrac{\text{Distance between Foci}}{\text{Major Axis}}$

**Figure 1.13** The parameters of an elliptical orbit.

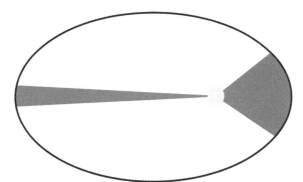

The radius vector (the line joining the centres of the Sun and Planet)
sweeps out equal area in equal times

**Figure 1.14** Kepler's Second Law.

It is worth pointing out the fundamental physics that lies behind the second law. A planet possesses both potential and kinetic energy. The potential energy relates to its distance from the Sun, reducing as it nears the Sun. The kinetic energy relates to its speed. As the planet is moving in space, there is no mechanism for it to lose energy, so the sum of the potential and kinetic energy must remain constant. In an elliptical orbit, the planet varies its distance from the Sun and so, for its total energy to be conserved, when its potential energy reduces as it nears the Sun its kinetic energy must increase. It must thus move faster along its orbit – exactly as implied by the second law.

The third law relates the period of the planet's orbit, $T$, with $a$, the semi-major axis of its orbit and states that:

**The square of the planet's period, $T$, is proportional to the cube of the semi-major axis of its orbit, $a$.**

One point should be made here: if an orbit is circular, then the semi-major axis is simply the radius of the circle and thus the distance of the planet from the Sun. When the orbit is not far from circular, then the semi-major axis is very close to the mean distance of the planet from the Sun. The third law is often stated in the form:

The square of the planet's period is proportional to the cube of its mean distance from the Sun.

However, this is not strictly accurate.

It should also be noted that Kepler's Third Law as stated above is only applicable when one of the bodies is significantly more massive than the other – as is always the case for the planets of our Solar System.

Writing the third law mathematically:

$$T^2 \, \alpha \, a^3$$

Thus:

$$T^2 = k \times a^3$$

where $k$ is a constant of proportionality.

The value of $k$, which will be the same for all objects orbiting the Sun, depends on the units chosen. It is conventional to measure the period, $T$, in units of Earth years and the semi-major axis, $a$, in units of the Earth's semi-major axis, which is termed an Astronomical Unit (AU), in which case $k = 1$.

The semi-major axis of the asteroid Ceres, which orbits the Sun every 4.60 years can thus be found using $T^2 = k \times a^3$. With $k = 1$, this becomes $a^3 = T^2$, and thus:

$$a = T^{2/3}$$
$$\text{giving } a = 2.77 \, \text{AU.}$$

Kepler's Third Law can, of course, be applied to any system of planets or satellites orbiting a body. Only the value of the constant of proportionality will be different.

### An example

Satellite television signals are broadcast from what are termed 'Geostationary' orbits above the equator. At a specific distance from the centre of the Earth a satellite will orbit the Earth once per day and so remain in the same position in the sky as seen from a location on the Earth's surface, thus allowing a fixed reception antenna. How high above the surface of the Earth at the equator would such an orbit be?

The radius of the Moon's orbit is 384 400 km and its orbital period is 27.32 days.

(You may be worried about this value for the period and might think that it should be 29.53 days. This latter value is the period between two New Moons, and is called the synodic lunar month. It is obviously related to the position of the Moon related to the Sun and so depends both on the Moon's motion about the Earth *and* the orbital motion of the Earth–Moon system around the Sun. The 27.32 day sidereal lunar month, the value that we need to use, is determined by the time it takes for the Moon to return to the same place on the celestial sphere after one orbit of the Earth. Incidentally, Richard Feynman in his famous *Feynman Lectures on Physics* made this mistake and used 29.5 days as the Moon's orbital period – but he was a world-famous physicist, not an astronomer, so perhaps he can be forgiven!)

Using these values, we can calculate the constant of proportionality that applies to satellites around the Earth:

$$k = (27.32)^2/(384\ 400)^3$$
$$= 1.314 \times 10^{-14}.$$

For our geostationary satellite, $T$ is 1, so we derive "'$a$' from:

$$1 = k \times a^3$$
$$a = (1/k)^{1/3}$$
$$= 42\ 377\ \text{km}.$$

The surface of the Earth is 6400 km from the centre, so geostationary satellites are ~36 000 km above the surface of the Earth.

## 1.8   Measuring the astronomical unit

A highly significant result of the third law is that it enabled astronomers to make a very accurate map of the Solar System. The relative positions of the planets could be plotted precisely, but the map had no scale. It was like having a very good map of a country but not knowing, for example, how many centimetres on the map related to 1 km on the ground. A way to solve this problem would be to make an accurate measurement of one reasonably large distance across the area covered

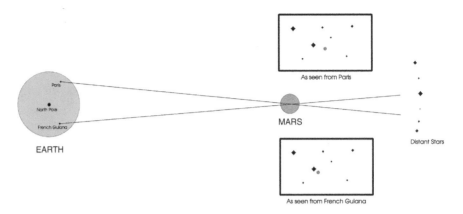

**Figure 1.15** Cassini's measurement of the distance to Mars.

by the map. This would then give the scale, and thus the distance between any other two points on the map could be found.

In the case of the Solar System, the obvious measurement to make was the distance between the Earth and either of its two nearest planets, Venus or Mars. Initially this was attempted by the use of parallax: the slight difference in direction of an object when viewed from different locations. In principle, the position of a planet, as seen against the backdrop of the distant stars, would be different when observed from separate locations on Earth. The problem is that the widest separation possible on the Earth is approximately its diameter, 12 756 km, which is small compared with the distances of the planets. The angular difference that has to be measured is thus very small and prone to errors.

In 1672, the Italian observer Cassini observed Mars from Paris whilst a colleague observed it from French Guiana in South America (Figure 1.15). They were able to measure its parallax, and hence measure the Earth–Mars distance. Using Kepler's Third Law they were then able to calculate the Earth's distance from the Sun, and derived a value of 140 million km.

Later astronomers observed the transit of Venus, two of which occur each century. By timing (from locations all over the Earth) when Venus first entered the Sun's limb and then just before it left, one can measure the parallax of Venus and hence find its distance. The value deduced from both the eighteenth century transits was 152.4 million km (Figure 1.16).

A truly accurate measurement had to wait until 1962, when powerful radars using large radio telescopes in the USA, Russia and at Jodrell Bank in the UK, were able to obtain echoes from the surface of Venus. The accepted value now is 149 597 870.691 km, just less than 150 million km or 93 million miles.

**Figure 1.16** The transit of Venus on June 8, 2004 seen through hazy cloud. Image: Ian Morison using a 150 mm refractor.

## 1.9   Isaac Newton and his Universal Law of Gravity

Isaac Newton was born in the manor house of Woolsthorpe, near Grantham, in 1642, the same year Galileo died. His father, also called Isaac Newton, had died before his birth and his mother, Hannah, married the minister of a nearby church when Isaac was 2 years old. Isaac was left in the care of his grandmother and effectively treated as an orphan. He attended the Free Grammar School in Grantham but showed little promise in academic work and his school reports described him as 'idle' and 'inattentive'. His mother later took Isaac away from school to manage her property and land, but he soon showed that he had little talent and no interest in managing an estate.

An uncle persuaded his mother that Isaac should prepare for entering university and so, in 1660, he was allowed to return to the Free Grammar School in Grantham to complete his school education. He lodged with the school's headmaster who gave Isaac private tuition and he was able to enter Trinity College, Cambridge in 1661 as a somewhat more mature student than most of his contemporaries. He received his bachelor's degree in April 1665 but then had to return home when the University was closed as a result of the Plague. It was there, in a period of 2 years and whilst he was still under 25, that his genius became apparent.

There is a story (which is probably apocryphal) that Newton was sitting under the apple tree in the garden of Woolsthorpe Manor. He might well have been able to see the first or last quarter Moon in the sky. It is said that an apple dropped on his head (or thudded to the ground beside him) and this made him wonder why the Moon did not fall towards the Earth as well.

Newton's moment of genius was to realise that the Moon *was* falling towards the Earth! He was aware of Galileo's work relating to the trajectories of projectiles and, in his great work *Principia* published in 1686, he considered what would happen if one fired a cannon ball horizontally from the top of a high mountain where air resistance could be ignored. The cannon ball would follow a parabolic path to the ground. As the cannon ball was fired with greater and greater velocity it would land further and further away from the mountain. As the landing point becomes further away the curvature of the Earth must be considered. In a more popular work published in the 1680s called *A Treatise of the System of the World*, he included Figure 1.17. The mountain is impossibly high in order for it to reach above the Earth's atmosphere. However, this is a thought experiment, not a real one. One can see from this that, if the velocity is gradually increased there would come a point when the cannon ball would never land – and would be in an orbit around the Earth.

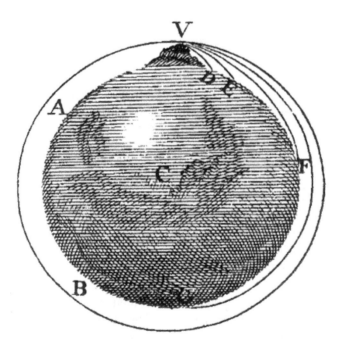

**Figure 1.17** Newton's thought experiment using a cannon ball.

Let us first treat this experiment quantitatively and use modern day values and units to extend Newton's arguments to encompass the Moon. You can calculate that the surface of the Earth falls below a flat horizontal line by approximately 5 m over a distance of 8 km. If one drops a mass from rest at the surface of the Earth, it will drop a height of $\frac{1}{2}gt^2$ in a time $t$, where $g$ is the acceleration due to gravity at the Earth's surface. The value of $g$ is $9.8\,\mathrm{m\,s^{-2}}$ so the fall would be 4.9 m. So, if we fired the cannon ball with a speed of about $8000\,\mathrm{m\,s^{-1}}$, its fall after 8 km would match the falling away of the Earth's surface and the cannon ball would remain in orbit.

Newton applied the same logic to the motion of the Moon, realising that, if the gravitational attraction between the Earth and Moon caused it to fall by just the right amount, it too would remain in orbit around the Earth. He knew enough about the Moon to be able to calculate the value of the acceleration of gravity at the distance of the Moon. For this, he needed to know the radius of the Moon's orbit (assumed to be circular) about the centre of the Earth, and also the period of its orbit around the Earth. Newton used a value of the radius of the Moon's orbit of 384 000 km and a period of 27.32 days or $2.36 \times 10^6\,\mathrm{s}$. Referring to Figure 1.18 (which is not to scale) one can easily calculate the distance, $L$ (the length AB), and direction (tangential to the radius vector) that the Moon would travel in 1 s if suddenly there were no gravitational attraction between the Earth and the Moon.

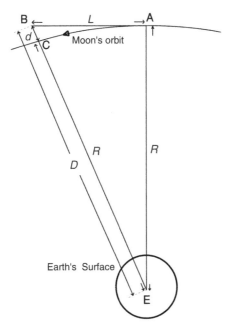

**Figure 1.18** Geometry of the Earth–Moon system.

Using the small angle approximation, where sin θ = θ (in radians), then $L = R \times \theta$, so

$$\theta = (1/2.36 \times 10^6) \times 2 \times \pi = 2.66 \times 10^{-6} \text{ rad}.$$

Giving        $L = 1.022 \text{ km}.$

As a result of the mutual attraction of the Earth and the Moon, the Moon will actually follow the curved path AC which can be thought of as being made up of the straight line motion AB and a fall BC, the distance fallen by the Moon in 1 s.

Let the distance from the centre of the Earth, point E, to A be R and from E to B be D. Finally let the distance from B to C be d. As the orbit is circular, the distance from E to C is also R.

With this notation, $d = D - R$
In the right-angled triangle ABE, $R/D = \cos \theta$,
or

$$D = R/\cos \theta.$$

Hence,

$$d = D - R = R/\cos \theta - R = R \left[(1/\cos \theta) - 1\right].$$

Now θ is a very small angle, so we may write:

$$\cos \theta = 1 - (\theta^2/2)$$

where θ is in radians.

Using the binomial theorem to find 1/cosθ, and ignoring all but the first two terms (θ is *very* small) we get:

$$1/\cos \theta = 1 + (\theta^2/2).$$

Substituting in the expression for BC, we obtain:

$$
\begin{aligned}
d &= R \times \theta^2/2 \\
&= [3.84 \times 10^8 \text{ m} \times (2.66 \times 10^{-6})^2]/2 \\
&= 1.36 \times 10^{-3} \text{ m}.
\end{aligned}
$$

So, in 1 s, the Moon falls vertically towards the Earth by just over 1 mm.

Let us assume that the acceleration due to gravity at the distance of the Moon is $g_m$, then this fall would be equal to $\frac{1}{2}g_m t^2$ so that $g_m$ is $0.00272\,\mathrm{m\,s^{-2}}$. This is considerably less than the value of 9.81 at the Earth's surface, so the force of gravity must fall off as the distance between the two objects increases. The value of $g$ at the distance of the Moon compared with that at the Earth's surface was $0.00272/9.81 = 2.77 \times 10^{-4}$. This is a ratio of $1/3606$. Newton knew that the radius of the Earth was 6400 km, so that the Moon, at a distance of 384 000 km, was precisely 60 times further away from the centre of the Earth than the Earth's surface. Hence, the value of $g$ at the distance of the Moon had fallen almost precisely by the ratio of the distances *from the centre of the Earth* squared!

This led Newton to his famous inverse square law: the force of gravitational attraction between two bodies decreases with increasing distance between them as the inverse of the square of that distance.

However, Newton had a problem: he felt that he could not publish his law until he could prove that the gravitational pull exerted by a spherical body was precisely the same as if all the mass were concentrated at its centre. This can only be proved by calculus and it took Newton a while to develop the ideas of calculus, which he called 'fluxions'. It was only then that he felt confident enough to present his theory to the world.

The proof is not difficult, but rather long: as a sphere can be regarded as set of thin nested shells, the required proof is to show that the gravitational attraction of a thin shell is as if all its mass were concentrated at its geometrical centre. You will thus see that there is no requirement that the body is uniformly dense (the Earth certainly is not) but it does require that the density at a given distance from the centre is constant – the body must have a spherically symmetric mass distribution.

Newton realized that the force of gravity must also be directly proportional to the object's mass. Also, based on his third law of motion, he knew that when the Earth exerts its gravitational force on an object, such as the Moon, that object must exert an equal and opposite force on the Earth. He thus reasoned that, due to this symmetry, the magnitude of the force of gravity must be proportional to both the masses.

His law thus stated that the force, $F$, between two bodies is directly proportional to the product of their masses and inversely proportional to the distance between their centres. This can be written as:

$$F \propto M_1 M_2 / d^2$$

where $M_1$ and $M_2$ are the masses of the two bodies and $d$ is their separation. Thus one can write:

$$F = G \times M_1 \, M_2 / d^2$$

which is Newton's Universal Law of Gravitation, where $G$ is the constant of proportionality called the universal constant of gravitation.

Why Universal? Using his second law (force = mass × acceleration) and his Law of Gravity, Newton was able to deduce Kepler's third law of planetary motion. This deduction showed him that his law was valid throughout the whole of the then known Solar System. To him that was Universal!

### 1.9.1    *Derivation of Kepler's third law*

To simplify the derivation, it is assumed that the orbit of the planet is circular.

The acceleration that must act on the object to keep it in a circular orbit around a body is given by:

$$a = v^2 / r$$

where $v$ is the object's speed in its orbit and $r$ is the radius of the circular orbit.

Also, stating Newton's Second Law:

$$F = m \, a$$

where $F$ is force, $m$ is mass and $a$ is acceleration.

So the inward force to act on the planet to overcome its inertia and keep it in a circular orbit is:

$$m_p v^2 / r.$$

This force is provided by the gravitational force between the Sun and the planet so:

$$m_p v^2 / r = G \, m_s m_p / r^2.$$

Cancelling $m_p$ and $r$, we get:

$$v^2 = G\, m_s/r$$

The period $P$ of the orbit is simply $2\pi r/v$, so $v = 2\pi r/P$.

Thus, $\qquad\qquad\qquad\qquad\qquad 4\pi^2 r^2/P^2 = G\, m_s/r$

Giving: $\qquad\qquad\qquad\qquad\qquad\quad 4\pi^2 r^3 = G\, m_s P^2$

Dividing both sides by $Gm_s$ and swapping sides gives:

$$P^2 = (4\pi^2/Gm_s)r^3.$$

As the part in brackets is a constant, $P^2$ is proportional to $r^3$ – Kepler's Third Law!

Over 300 years after the publication of his great theory, it is difficult to be precise as to how he developed the theory in his mind. The outline above is, I am sure, how Newton would like us to believe it happened – observations leading to an inverse square law that was proven by showing that Kepler's Third Law could be derived from it. However, it could have happened in an inverse manner. From Newton's Second Law and Kepler's Third Law one can show that the forces between the Sun and planets must follow an inverse square law. From this, Newton could calculate the value of $g$ at the distance of the Moon, and hence the period of its orbit by working backwards through the calculation above. This observation would then show that a prediction of the law was true. It does not really matter which was the case, Newton's Universal Law of Gravitation was an outstanding achievement.

Newton derived a value of $G$ by estimating the mass of the Earth assuming it had an average density of $5400\,\mathrm{kg\,m^{-3}}$. He suspected that the Earth increases in density with increasing depth and had simply doubled the value of $2700\,\mathrm{kg\,m^{-3}}$ that is measured at the surface of the Earth. (This was a pretty good – and very lucky – estimate, as it is actually $5520\,\mathrm{kg\,m^{-3}}$!)

Again, using modern values and units, let us carry out this calculation.

We derived a value for the acceleration due to gravity at the distance of the Moon, $g_m$, above:

$$g_m = 0.00272\,\mathrm{m\,s^{-2}}.$$

From Newton's Second Law (force = mass × acceleration), the force acting on the Moon must have been:

$$M_m \times 0.00272 \, \text{kg m s}^{-2} = M_m \times 0.00272 \, \text{N}$$

Equating this with the force between them, as calculated from his law of gravity:

$$G \times M_e M_m / d^2 = M_m \times 0.00272,$$

Giving

$$G = 0.00272 \times d^2 / M_e$$

(Notice that the Moon's mass cancels out.)

With

$$d = 384\,000 \, \text{km} = 3.84 \times 10^8 \, \text{m}$$

and

$$M_e = 5400 \times 4/3 \times \pi \times (6.4 \times 10^6)^3 \, \text{kg}$$
$$= 5.93 \times 10^{24} \, \text{kg}$$
$$G = 0.00272 \times (3.84 \times 10^8)^2 / 5.93 \times 10^{24} \, \text{N m}^2 \text{kg}^{-2}$$
$$= 4.0 \times 10^{14} / 5.93 \times 10^{24} \, \text{N m}^2 \text{kg}^{-2}$$
$$= 6.76 \times 10^{-11} \, \text{N m}^2 \text{kg}^{-2}$$

Due to his lucky estimate of the mean density of the Earth, this was a very good result – the now accepted value of $G$ being $6.67 \times 10^{-11} \, \text{N m}^2 \text{kg}^{-2}$.

## 1.10    Experimental measurements of G, the universal constant of gravitation

In 1774, Nevil Maskelyne used the deflection (about 11 arcsec) of a plumb line, on the slopes of Mt Schiehallion in Scotland, to determine the gravitational attraction between the plum bob and the mountain. He was primarily interested in using this result to measure the mean density of the Earth. Schiehallion, which rises to 3547 ft (~1081 m), has a very regular shape, and so Maskelyne was able to estimate the mountain's mass and thus determine the value of $G$, but his values for $G$ and that of the density of the Earth (4400 kg m$^{-3}$) were not very accurate.

Later, in 1798, Henry Cavendish was the first to measure $G$ in the laboratory using equipment devised by John Mitchell, who had died before being able to complete the experiment. Mitchell had built a very sensitive torsion balance (Figure 1.19) for the measurement: a horizontal beam with small lead balls at each end is suspended from its centre by a thin torsion wire. Large lead balls

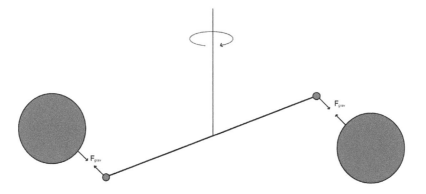

**Figure 1.19** The Cavendish Experiment.

were then placed near each of the small balls in the same horizontal plane such that the gravitational attraction of the large balls on the small balls tended to twist the torsion wire in the same direction. The force due to the gravitational attraction between the ball pairs was balanced by the force produced by the twisting of the torsion wire. In a further experiment, the force required to twist the wire was measured by timing the free oscillations of the beam about the axis of the torsion wire. From his measurements, Cavendish derived a value for $G$ of $6.75 \times 10^{-11} \, \mathrm{N\,m^2\,kg^{-2}}$.

**1.11**  ## Gravity today: Einstein's special and general theories of relativity

In 1905, as mentioned in Section 1.6.6, Albert Einstein published his paper on the Special Theory of Relativity. One consequence of this theory was that nothing could travel *through* space faster than the speed of light. This extended beyond material things: information and the effects of force fields were included too. Einstein realised that this was not compatible with Newton's Law of Gravitation which implied 'instantaneous action at a distance'.

I have italicized the word 'through'. It *is* possible for the expansion of space to carry objects apart faster than the speed of light – indeed we believe that this happened during a period called 'inflation' close to the origin of the universe. Imagine making a currant bun: currants are packed tightly into the dough which is then placed into an oven. When cooked, one hopes that the dough will have expanded, increasing the separation of the currants. They have not moved *through* the dough, but have been carried apart by the *expansion* of the dough.

Perhaps a thought experiment will help to make the incompatibility of Newton's theory with special relativity clear. Suppose that the Sun could suddenly cease to exist. Under Newton's gravity, where the effect of a change in mass would be instantly felt throughout the universe, the Earth would immediately fly off at a tangent. Einstein realised that this could not be the case. Not only would we not be aware of the demise of the Sun for 8.31 min – the time that light takes to travel from the Sun to the Earth – but the Earth must continue to feel the gravitational effects of the Sun for just the same time, and would only fly off at a tangent at the moment we ceased to see the Sun. This assumes, of course, that whatever carries the information about the gravitational field of the Sun will also propagate at the speed of light.

Consequently, something has to propagate through space to carry the information about a change in gravity field. Einstein thus postulated the existence of gravitational waves that carry such information. As we will see in a later chapter, the existence of such gravitational waves has already been shown indirectly and it is likely that, before long, direct evidence of their existence will be gained.

This concept came out of Einstein's 1915 paper on 'The General Theory of Relativity'. (Please note: it is not, as is often stated, 'The Theory of General Relativity' although in shorthand it is often referred to correctly as 'General Relativity'.) In essence, this is a relativistic theory of gravity in which gravity is a concept that we infer to explain what we observe. General relativity regards that the motion of objects through space *not* in a straight line is due, not to a force applied to it as in Newtonian gravity, but is its natural path in space which has been 'curved' due to the presence of mass. (To be precise, we should use the term space–time rather than space, but this is an introductory text!)

In Einstein's view, in the absence of mass, space is flat. This term (which the author does not like as it implies that space is two-dimensional) actually means that within 'flat space' Euclidean geometry holds true. That is, if one lays out a large triangle in any orientation in space, the inscribed angles will add up to 180°. A second property of 'flat space' is that two initially parallel laser beams separated by some distance will remain parallel.

If, now, mass is introduced into flat space, the space around it becomes curved and light or matter will follow curved, rather than straight trajectories. An analogy can help. Imagine a stretched horizontal flat rubber sheet. Ball bearings will roll across it in straight lines. If a lead ball is now laid on the sheet it will cause a depression (Figure 1.20) and, if ball bearings now come close to the depression, they will follow curved lines. In Einstein's view, the orbital motion of the Earth around the Sun is simply the natural motion of a body through curved space – the force we perceive is actually just a consequence of curved space–time.

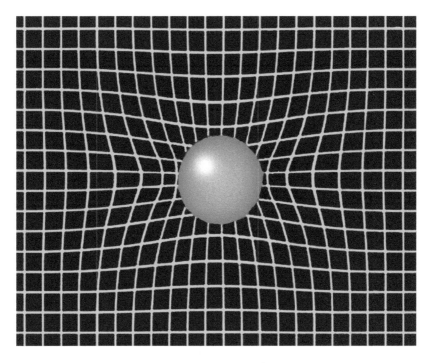

**Figure 1.20**  A rubber sheet analogy of the distortion of space–time by mass.

Let us consider a further 'thought experiment'. Imagine that the northern hemisphere of a large planet had a totally frictionless surface. The inhabitants have an unshakeable belief that the surface of their planet is flat. Two explorers set off simultaneously on sledges from two points, 10 km apart, on the equator. They travel on parallel paths pointing true north with identical initial velocities. As there is no friction, they will continue at the same speed and would expect to remain 10 km apart. They will thus be somewhat surprised (and possibly hurt) when they collide at the North Pole! They might try and explain how this could have happened without giving up their belief that the surface of their planet is flat. The only way that they could do this would be to postulate a force – that they might call *gravity* – that attracted them towards each other.

**In the same way, gravity is a force that *we* infer to explain what happens in a curved three-dimensional space if we assume that space is flat, not curved.**

## 1.12    Conclusion

This chapter has brought us from the middle of the sixteenth century to the start of the twentieth century and picked out some key advances in our understanding of the Universe. Several concepts described in this chapter will be used later in the book:

- The observed apparent magnitude of a star is one requirement for the calculation of its luminosity.
- Accurate measurement of time is vital in the study of pulsars which are the rotating remnants of supernova explosions.
- When we consider binary star systems, we will use a form of Kepler's Third Law, as modified by Newton, to carry out calculations where the two orbiting masses are comparable.
- Newton's Law of Gravity and the measured value of G, the universal constant of gravitation, will be used to calculate the mass of the Sun.
- Without a precise knowledge of planetary orbits and the scale size of the Solar System as determined by the measurement of the Astronomical Unit, it would have been impossible to send spacecraft to orbit and land on the Moon and planets, and our knowledge of the Solar System would be greatly limited.
- Einstein's Theory of General Relativity underpins the whole subject of Cosmology – the evolution of the Universe.

The astronomers and physicists whose work is described in this chapter have thus given a very solid foundation on which modern astronomy and our understanding of the Universe is built.

## 1.13    Questions

1.  How much brighter will a star of 1st magnitude appear compared with one of 9th magnitude?
2.  What is the apparent magnitude of a star that appears 251 times less bright than the 0th magnitude star Vega?
3.  An observer whose dark-adapted pupil is 5 mm in diameter can observe a star of 6th magnitude towards the zenith (overhead). He now observes with a pair of binoculars having an objective size of 40 mm. The sensitivity of the binoculars compared with the eye is proportional to the ratio of the areas of the objective and pupil so that, using the binoculars, the observer

will be able to observe fainter objects. What is the faintest magnitude star that the observer might hope to observe at the zenith with their use?

4.  The observed brightness of a luminous object falls off as the square of its distance from us (the inverse square law). Two stars of identical luminosity are observed: the apparent brightness of the further one is 10 000 times less than the nearer.

    (a)  What is the difference in apparent magnitude?

    (b)  How many times further away is it?

5.  A star is observed to cross the meridian at an elevation of $67°$, as seen from an observatory at a latitude of $52°$ north. What is the declination of the star?

    What would be the declination of a star observed to transit at an elevation of $20°$?

6.  A star is observed to cross the meridian (due south) at an elevation of $34°$, as seen from an observatory sited at a latitude of $42°$ north. What is the declination of the star? At the moment of transit, a clock running on Universal Time (UT) read 03 h 16 min 24 s. At the previous midnight, the sidereal time was 14 h 38 min 54 s. Calculate the Right Ascension of the star.

7.  The newly discovered 'dwarf planet', Eris, has an elliptical orbit with a semi-major axis of $67.89$ AU. Calculate its period in years.

8.  An asteroid is orbiting the Sun at a distance of $2.7$ AU. How long does it take to orbit the Sun?

9.  Assuming that Venus has a circular orbit and that it orbits the Sun every $224.7$ days, calculate its distance from the Sun in Astronomical Units.

    As Venus passed closest to the Earth a planetary radar at Jodrell Bank measured the round trip travel time of a radar pulse to be $272$ s. Estimate the value of the Astronomical Unit, the semi-major axis of the Earth's orbit around the Sun. (Assume that the Earth also a circular orbit and that the velocity of light is $3 \times 10^8 \, \mathrm{m \, s^{-1}}$.)

## Chapter 2

# Our Solar System 1 – The Sun

In this chapter we will cover how the Solar System was formed and study our Sun: how it generates its energy by nuclear fusion, the characteristics of its atmosphere and how its solar wind and magnetic field interacts with our Earth. Finally, to continue the story of Einstein's theories in Chapter 1, we will see how solar eclipses were able to provide proof of Einstein's General Theory of Relativity – his theory of gravity.

## 2.1 The formation of the solar system

It is believed from studies of meteorites – remnants of the early Solar System – that our Sun and planets formed some 4.6 billion years ago from what is called a giant molecular cloud. The gas and dust making up this cloud had been created over billions of years by the processing of primeval hydrogen and helium in stars to create heavier elements that are then ejected into space at the end of their lives.

The nebula hypothesis, which explains how the Sun and planets were created, was first proposed by Emanuel Swedenborg in 1734 and then, independently, by Pierre-Simon Laplace in 1796. It is now thought that the initial cloud of dust and gas would have been about 1 pc (3.26 Ly) across and that its collapse was triggered by the shock waves from one or more supernovae (the explosive end to giant stars). These would have produced regions of higher density than their surroundings which would then collapse under their own gravity. The region that spawned our Solar System would have had a diameter of perhaps 13 000 AU with a total mass perhaps twice that of our Sun. Its composition would have been similar to that of our Sun, with 98% made up of hydrogen (~74%) and helium (~24%) and about 2% of heavier elements.

It is interesting to get a feel of its composition from a list of some of the most common isotopes of the elements in nuclei per million:

| Hydrogen-1 | 705 700 |
|---|---|
| Helium-4 | 275 200 |
| Carbon-12 | 3032 |
| Nitrogen-14 | 1105 |
| Oxygen-16 | 5920 |
| Neon-20 | 1548 |
| Sodium-23 | 33 |
| Magnesium-24 | 513 |
| Aluminum-27 | 58 |
| Silicon-28 | 653 |
| Sulphur-32 | 396 |
| Argon-36 | 77 |
| Calcium-40 | 60 |
| Iron-56 | 1169 |
| Nickel-58 | 49 |

Notice that those elements whose nuclei have an integral multiple of four nucleons are relatively common. We will see when we study stellar evolution that these elements are those produced by the build up of helium nuclei and thus their atomic numbers are integral multiples of four. Iron is also common as it is the element with the most stable nucleus and the virtual end point of the nuclear fusion processes during the lifetime of a star. Elements of higher atomic number are only created when a star explodes at the end of its life and are thus comparatively rare.

It is likely that the solar nebula would have had some rotational energy. As the nebula collapsed, conservation of angular momentum meant that it would have begun to spin faster and, as it became denser, collisions within it would cause the gas to heat up. This would increase the pressure within the nebula so tending to make it expand and thus preventing further collapse. The term 'Giant Molecular Cloud' used for the initial gas and dust cloud from which our Solar System formed implies that it contained many molecules. These played an important role in allowing the nebula to condense in that transitions between their vibrational states can emit long wavelength infrared photons. These could escape through the dust and so carry energy away from the cloud so preventing a build up of heat that would have prevented further collapse.

Sir James Jeans showed that a nebula would only collapse if it was sufficiently massive. A gas cloud would have to exceed what is termed the Jeans Mass in order to collapse – a dense, cool cloud being able to collapse more easily than a less dense, warmer cloud. The initial mass required, even for a dense, cool cloud, has

**Figure 2.1** The Pleiades cluster. Image: Digitized Sky Survey NASA/ESA/AURA/Caltech.

to be many solar masses so it is virtually impossible for a single star to form on its own and so young stars are seen in clusters like the Hyades and Pleiades clusters in the constellation Taurus.

The Pleiades cluster, shown in Figure 2.1, contains about 500 stars. It is believed that the Pleiades cluster is passing through a dust cloud and this gives rise to the scattered light that is reflected from regions surrounding the brighter stars – known as a reflection nebula.

Due to the forces of gravity and the net angular momentum of the nebula, the nebula flattened into a spinning protoplanetary disc. Its overall diameter was roughly 200 AU with a dense central region that rapidly increased in temperature to form what is called a protostar (Figure 2.2).

After around 100 million years, the temperature and pressure at the core of the protostar became so great (~10 million K) that its hydrogen began to fuse into helium and the pressure produced by the resultant gamma rays became able to counter the force of gravitation. The fledgling star went through a turbulent phase, throwing off perhaps half its mass, until it finally stabilized. The protostar had become a star, our Sun.

From the gas and dust surrounding the nascent star, the various planets were formed through a process known as accretion. Dust grains in orbit around the protostar clumped together and formed what are called planetesimals, between 1 km and 10 km in diameter, which then gradually increased in size by further

**Figure 2.2**  Artist's impression of the solar nebula. Image: Ames Research Centre, NASA.

collisions. Due to the Sun's radiation, the inner Solar System was too warm for volatile molecules like water and methane to condense, so the planetesimals which formed there were relatively small and composed largely of compounds with high melting points, such as silicates and metals. These rocky bodies eventually became the terrestrial planets: Mercury, Venus, Earth and Mars. As a result of the gravitational effects of Jupiter, the formation of a planet between Mars and Jupiter was disrupted leaving rocky objects that are known as minor planets or asteroids in what is called the asteroid belt. The largest of these, Ceres, has recently been given the status of a 'dwarf planet'.

As the temperature fell further away from the Sun, volatile icy compounds could remain solid – beyond what is called the frost line. Jupiter and Saturn were able to gather far more material than the terrestrial planets and overlaying their icy/rocky cores were layers of metallic and molecular hydrogen. They became the gas giants and contain the largest percentages of hydrogen and helium. Uranus and Neptune captured much less material and are known as ice giants because their cores are believed to be mostly made of ice which is overlain by molecular hydrogen and gases such as ammonia, methane, and carbon monoxide.

As our young Sun settled down as a stable hydrogen burning star it went through a phase – called the T-Tauri phase – when there was a major outflow of material 'boiling off' from its surface. This outflow still continues, but at a far slower rate, and is called the solar wind. As a result the protostar lost much of its original mass. This strong solar wind cleared away all the remaining gas and dust in the protoplanetary disc into interstellar space, thus ending the growth of the planets.

The majority of the moons probably formed at the same time as their parent planets. However, it seems that our own Moon probably formed later when a body several times as massive as Mars collided with our planet. The giant impact blasted molten rock into orbit around Earth which then cooled to form the Moon.

From radiometric dating, we believe that the oldest rocks on Earth are approximately 3.9 billion years old. We expect that the Solar System is older than this as the Earth's surface is constantly evolving as the result of erosion, volcanism and plate tectonics. It is believed that meteorites were formed early on within the solar nebula so estimates of their age should give us an age of the Solar System. The oldest meteorites are found to have an age of ~4.6 billion years, giving us a minimum age of the Solar System.

## 2.2    The Sun

Our Sun is certainly a typical star; its structure and the nuclear fusion process by which it generates its energy are applicable to the vast majority of stars. However, it is not, as is often stated, an average star. As we will see in Chapter 6, there are seven stellar types: ranging from the hottest to the coolest these are Types O, B, A, F, G, K, and M. Our Sun is a Type G star and thus appears, if anything, to be below average. This ignores the fact that there are far more cool stars than hot ones. Our Sun is of Type G2 (at the hotter end of the G type stars), and ~95% of all stars are smaller and cooler than the Sun. Our Sun is well above average!

### 2.2.1    *Overall properties of the Sun*

Observations from Earth, coupled with the basic laws of physics, enable us to measure the main properties of the Sun.

#### Diameter

The Sun subtends an angle of ~30 arcmin in the sky. Given its distance from us of ~150 000 000 km one can directly estimate the Sun's diameter:

$$D = R\theta \text{ (With } \theta \text{ in radians and R is the Earth-Sun distance.)}$$

$$\theta = 30/(60 \times 57.3)\,\text{rad}$$
$$= 8.7 \times 10^{-3}\,\text{rad}$$

So
$$D = 1.5 \times 10^8 \times 8.7 \times 10^{-3}\,\text{km}$$
$$= 1\,308\,900\,\text{km}$$

This is reasonably close to the accurate value of 1 391 978 km. The Sun's radius is thus ~700 000 km.

(The distance of the Earth from the Sun varies from a minimum of 147 085 800 km when closest to the Sun on January 3 to 152 104 980 km when furthest from the Sun on July 4. To derive a precise value one would measure the observed angular diameter when at a known distance from the Sun.)

## Mass

The mass of the Sun can be derived using the same methods that Newton employed with the Moon. Equating the force attracting the Earth to the Sun given by Newton's Law of Gravitation with that due to centripetal acceleration we get:

$$MmG/R^2 = mv^2/R$$

where $M$ is the mass of the Sun, $m$ is the mass of the Earth, $v$ is the velocity of the Earth around the Sun, $G$ is the universal constant of gravitation and $R$ is the (assumed constant) distance of the Earth from the Sun.

The mass of the Earth, $m$, cancels out, giving:

$$M = v^2 R/G$$

But $v = 2\pi R/P$ where $P$ is the period of the Earth's orbit, so substituting:

$$M = 4\pi^2 R^3/GP^2$$
$$= 4 \times (3.14159)^2 \times (1.496 \times 10^{11})^3/6.67 \times 10^{-11} \times (3.156 \times 10^7)^2\,\text{kg}$$
$$= 1.99 \times 10^{30}\,\text{kg}$$

The mass of the Sun is $2 \times 10^{30}\,\text{kg}$.

## Density

From the radius and mass of the Sun one can easily derive its average density:

$$\text{Volume} = 4/3\pi r^3$$
$$= 4/3\pi(700\,000\,000)^3$$
$$= 1.4 \times 10^{27}\,\text{m}^3$$

So the density is:

$$= M/V$$
$$= 2 \times 10^{30}/1.4 \times 10^{27}\,\text{kg m}^{-3}$$
$$= 1428\,\text{kg m}^{-3}$$

This is about 40% greater than that of water and about 26% that of the Earth.

## 2.2.2     *The Sun's total energy output*

It might surprise you that any of you could almost certainly derive a value of the Sun's energy output good to a factor of 10 simply by knowing the distance of the Sun and using your own experience!

Let us assume one has a value, $e$, for the Sun's energy falling on $1\,\text{m}^2$ of the Earth's atmosphere. This is just $1\,\text{m}^2$ of a spherical surface centred on the Sun which has an area of:

$$A = 4\pi(1.5 \times 10^{11})^2\,\text{m}^2$$

where $1.5 \times 10^{11}\,\text{m}$ is the distance of the Earth from the Sun. As all of the Sun's energy must pass through this surface, the Sun's total energy output is simply $A \times e$.

What is the value of $e$? First imagine that you were standing in a desert directly under the Sun in a cloudless sky. Then instead, imagine standing in a room under a 100 W bulb, a 1000 W infrared heat lamp or a 10 kW arc lamp. Which might be comparable? I suspect that you would pick the heat lamp. So a 'guestimate' of the power from the Sun falling on $1\,\text{m}^2$ might well be 1000 W. This would actually be quite close as the measured value is 1370 W.

The Sun's total energy output is thus:

$$E = 1370 \times 4\pi(1.5 \times 10^{11})^2\,\text{W}$$
$$= 3.86 \times 10^{26}\,\text{W}$$

### 2.2.3    *Black body radiation and the sun's surface temperature*

In order to estimate the surface temperature of the Sun it is necessary to know about the concept of black body radiation which arose out of quantum mechanics in the early years of the last century. A black body, a term introduced by Gustav Kirchhoff in 1860, is an object that absorbs all electromagnetic radiation that falls onto it – it neither reflects any radiation nor allows any radiation to be transmitted through it. One way of approximating a black body is to make a cavity with a very small aperture whose interior surface is matt black. Any radiation that enters the aperture will almost certainly be absorbed within the cavity. If the cavity were heated to some temperature then the small aperture will emit electromagnetic radiation and this is said to be black body radiation. Another name for this radiation is, not surprisingly, cavity radiation. As the radiation inside the cavity will be in thermal equilibrium with its walls, this will be a source of thermal radiation.

Thermal radiation has the property that both the emitted power and the wavelength at which the electromagnetic radiation is a maximum are directly related to its effective temperature; in the case of a cavity, this is its internal temperature. The radiation is then said to have a black body spectrum (Figure 2.3). Below about 700 K (430°C) black bodies produce very little radiation at visible wavelengths and appear black to our eyes – though we could sense the infrared radiation emitted at temperatures somewhat below this. Above this temperature

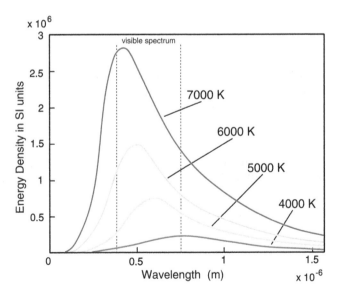

**Figure 2.3** Black body radiation curves.

black bodies will emit enough radiation at visible wavelengths for us to see a colour which passes through red, orange, yellow, and white to blue as the temperature increases. As the temperature increases further the peak wavelength moves into the ultraviolet, but there is still considerable energy in the blue part of the spectrum so that the object appears blue.

The problem of calculating the form of the spectral curve for black body radiation was finally solved in 1901 by Max Planck and is known as Planck's law of black body radiation. Planck had found a mathematical formula that fitted the experimental data, but to find a physical interpretation for this formula he had to invoke the principle of quantization: 'photons' within the cavity could only have certain allowed energies. In 1905, Einstein proposed that electromagnetic radiation was quantized in order to explain the photoelectric effect and was later awarded the Nobel Prize in Physics for this insight.

There are two laws which relate to electromagnetic radiation which follows a black body spectrum.

### Wien's Law

The first law is Wien's displacement law which states that the wavelength at which the spectrum peaks is inversely proportional to the surface temperature. (In other words, the *greater* the temperature, the *shorter* the peak wavelength.) The peak wavelength, $\lambda_{max}$, is thus given by a constant, Wien's constant, divided by the temperature. It value is $2.897 \times 10^{-3}$ m K so with $\lambda$ in metres and $T$ in kelvin:

$$\lambda_{max} = 2.897 \times 10^{-3}/T$$

or, with $\lambda$ in nanometres:

$$\lambda_{max} = 2.897 \times 10^{6}/T$$

Wien's Law enables the temperature of inaccessible objects, such as a lava flow or the interior of a blast furnace (an excellent black body as it really is a cavity), to be measured and, as we will see, allows us to estimate the surface temperature of the Sun.

### Stefan–Boltzmann Law

The second law relating to black body radiation is called the Stefan–Boltzmann Law, which is often known simply as Stefan's Law. It was discovered experimentally by Jožef Stefan in 1879 and derived theoretically from thermodynamic principles by Ludwig Boltzmann in 1884. The Law states that the total energy radiated per unit surface area of a black body per unit time is directly proportional to the fourth power of the black body's absolute temperature. The constant of proportionality, $\sigma$, is called the Stefan–Boltzmann constant or Stefan's constant, and has the value:

$$\sigma = 5.67 \times 10^{-8}\,\mathrm{Wm^{-2}K^{-4}}$$

Thus the emitted power across the electromagnetic spectrum of a body of surface area $A$ and temperature $T$ is given by:

$$E = \sigma AT^4\,\mathrm{W}$$

Real objects do not, in general, act as perfect black bodies but in the majority of astronomical applications we first assume that the object (such as a star or a planet) does act as a black body and then make a suitable correction if necessary.

An interesting astronomical result that came out of the radiation laws was related to the planet Mercury. It was long thought that Mercury's rotation period was the same at its orbital period of 88 days. If this was the case, one face would be locked to the Sun (like the Moon to the Earth) and would thus be very hot, but the face away from the Sun would be very cold – close to absolute zero. However, observations made at wavelengths of 11.3 and 1.9 cm in the radio part of the spectrum indicated that the supposed dark side surface temperature must be in the region of 250 K. This observation was later confirmed when radar observations showed that Mercury's rotation period was not 87.97 days, but 68.65 days, so all parts of the surface do, at some time, face the Sun.

Both the peak wavelength and total power output of a black body are related to its surface temperature. Assuming that the Sun's visible surface acts as a black body, this then gives us two ways of estimating its surface temperature.

The peak of the visible spectrum is at a wavelength of ~500 nm ($0.5 \times 10^{-6}\,\mathrm{m}$). Using Wien's displacement law, temperature is given by:

$$\begin{aligned} T &= 2.9 \times 10^6/\lambda_{peak}\,\mathrm{K} \quad (\text{where } \lambda_{peak} \text{ is in nanometres}) \\ &= 2.9 \times 10^6/500\,\mathrm{K} \\ &= 5800\,\mathrm{K} \end{aligned}$$

Using the total solar power output and the radius of the Sun (from the diameter as determined above) we can also use the Stefan–Boltzmann Law:

$$E = \sigma AT^4$$
$$= 5.671 \times 10^{-8} \times 4 \times \pi \times (6.95 \times 10^8)^2 \times T^4$$

So
$$T = \{4 \times 10^{26}/[5.671 \times 10^{-8} \times 4 \times \pi \times (6.95 \times 10^8)^2]\}^{-4}$$
$$= (4 \times 10^{26}/3.44 \times 10^{11})^{-4}$$
$$= 5839\,K$$

These agree quite well, but are a little higher than the accepted value of 5780 K.
So, in summary, the main properties of the Sun are:

| | |
|---|---|
| Diameter | $= 1391978\,km$ |
| Mass | $= 2 \times 10^{30}\,kg$ |
| Density | $= 1400\,kgm^{-3}$ |
| Luminosity | $= 3.86 \times 10^{26}\,W$ |
| Surface Temperature | $= 5780\,K$ |

## 2.2.4    *The Fraunhofer lines in the solar spectrum and the composition of the sun*

In 1666, Isaac Newton, using a prism, showed that sunlight is composed of all the colours of the spectrum and in 1804, William Wollaston observed that there appeared to be some gaps in the spectrum that looked like dark lines. Later, in 1911, Joseph Fraunhofer mapped many of these lines with reasonable accuracy. They have thus become known as Fraunhofer lines (Figure 2.4). They represent wavelengths where there is a lessening of the observed solar emission and are thus called absorption lines. Later, Gustav Kirchoff and Robert Bunsen found that the wavelengths of the absorption lines seen in the Sun corresponded to those of the emission lines observed when the atoms of a particular element are excited. This can be achieved by sprinkling a compound of the element into a Bunsen burner flame, when, for example, salt gives an orange colour due to a close pair of emission lines called the sodium D lines.

**Figure 2.4** The solar spectrum showing the Fraunhofer lines. The peak intensity is in the yellow part of the spectrum close to the strong pair of sodium D lines in the centre of the spectrum.

Before long, Fraunhofer lines corresponding to all the known elements had been found in the Sun's spectrum except for one set of lines. It was realised that there must be an element in the Sun's atmosphere that had not then been discovered on Earth. It was thus called helium after 'Helios' the Greek name for the Sun.

How are these lines formed? The photosphere will emit a continuous (almost black body) spectrum. The photons will then pass through the Sun's upper atmosphere, the chromosphere, where atoms can absorb photons that correspond to transitions between their energy levels. Thus the lines, called absorption lines, will be at just the same wavelengths as the emission lines that we can observe on Earth.

This is where many books stop, but it cannot be quite as simple as this. Within a short while all the atoms would be in their upper energy states and the absorption would stop! In a steady state situation, the atoms must eventually return to their original states and may do so by emitting the very same wavelengths as had been absorbed – which would imply that there would be no net absorption! The atoms will then emit in random directions so only a small percentage of the re-radiated emission from the line of sight atoms will come in our direction. However, atoms over the whole face of the Sun will also be emitting their photons in random directions. Their emission towards us will partly balance the emission that we do not receive from the atoms in our line of sight.

There is a second mechanism, called collisional de-excitation, which can prevent re-emission. Before the atoms have a chance to re-emit a photon they may interact with another atom in the atmosphere and the excitation energy is turned into kinetic energy of the atoms, so heating up the Sun's atmosphere.

As we will see in later chapters, observations of the lines in the spectra of stars and galaxies play a critical role in our understanding of both our Galaxy and the Universe.

From the analysis of the solar spectrum it is possible to estimate the composition of the majority of the Sun's interior as the outer layers are 'mixed' by convective currents as will be described later. The composition is about 71% by mass of hydrogen (91.2% in number of atoms), 27.1% by mass of helium (8.7% in number of atoms), oxygen makes up 0.97% (0.078% in number of atoms), and carbon 0.40% (0.043% in number of atoms). The small remainder then comprises all the other atoms detected in the Sun's spectrum. In the core there will be less hydrogen and more helium due to the nuclear fusion processes to be described below.

## 2.3   Nuclear fusion

Up to the late 1800s, scientists could not understand how the Sun could create so much light and heat. If the Sun had been entirely made of something like coal (along with the oxygen it would need to burn) it would burn itself up in about a thousand years! Since the Sun had been providing heat and light for at least several thousand years a chemical source of the Sun's energy was clearly impossible. Around 1870, Hermann von Helmholtz realised that if the Sun was contracting in size, energy, derived from potential energy, could be released. He knew, from the derivations given above, the mass and size of the Sun and also knew how much energy the Sun is continuously creating and sending out into space. He calculated how much the Sun would have to reduce in size to provide its observed output, and deduced the Sun would be able to sustain its energy output for around 20 million years. In the late 1800s people were happy to assume that the Solar System was less than 20 million years old, so his idea was almost universally accepted as the likely way that the Sun creates its energy. (Note: Jupiter is radiating into space almost twice as much energy as it receives from the Sun – the excess comes because Jupiter is slowly contracting in size and so potential energy is being converted into heat.)

However, during the late 1800s, geologists established that many Earth rocks and the fossils within them are definitely millions of years old, so Helmholtz must have been wrong. In 1905, Einstein published the famous $E = mc^2$ equation as part of his Special Theory of Relativity. One could thus surmise that, as the velocity of light, $c$, is very large, a tiny amount of mass ($m$) might be converted into an enormous amount of energy ($E$). By around 1925, physicists had determined the mass of a proton (a hydrogen nucleus) and had also determined that an alpha particle (a helium nucleus) has a mass slightly less than that of four protons. They realised that four hydrogen nuclei might be able to 'fuse' together into a helium nucleus (called fusion) and the mass that apparently disappears could be converted into energy.

This is difficult! Since the hydrogen nuclei are each positively charged protons, they repel each other. In order to overcome this mutual repulsion, the protons must be moving toward each other nearly at the speed of light – and even then quantum mechanical tunnelling must be invoked. They can only do this if it is very hot – of the order of 10 million K. Due to the great mass of the Sun, the pressure at its centre (called its core) must be very high to oppose the mass of the overlaying layers that form the greater part of the Sun and calculations show that the core would reach and exceed the required temperature. Thus the source of the Sun's energy is a nuclear fusion reactor within the core of the Sun. As the dust and gas that made up our Sun collapsed under gravity, the temperature at the centre increased and the protons began to move faster. When the temperature exceeded ~10 million K, the protons' kinetic energy became sufficient for two protons on a collision course to get sufficiently close to allow an effect called quantum mechanical tunnelling to come into play. This allows one of them to overcome the potential barrier due to the electrostatic force between them.

In quantum theory, particles, such as the two protons approaching each other, can be described by wave functions, which represent the probability of finding a particle in a certain location. If a particle is adjacent to a potential barrier, its wave function decays exponentially through the barrier, but will have still have a very small amplitude on the far side of the barrier. There is thus a very small probability that the particle can 'appear' on the other side of the barrier in which case it is then said to have tunnelled through it. Quantum tunnelling thus allows a particle, in this case a proton, to violate the principles of classical mechanics by passing through a potential barrier higher than the kinetic energy of the particle.

This then (very rarely) allows the two protons to come sufficiently close for the strong nuclear force to momentarily bind them together before one of the protons decays into a neutron, a positron and an electron neutrino (see box). This leaves a deuteron – the combination of a proton and a neutron which is the nucleus of deuterium. The positron then annihilates with an electron, and their mass energy is carried off by two (sometimes more) gamma ray photons. This is the first step in what is called the proton–proton cycle outlined below.

**Positron**

A positron is the antimatter counterpart of the electron. It has an electric charge of $+1$, a spin of $1/2$, and the same mass as an electron. When a positron collides with an electron, annihilation occurs, resulting in the production of two (or more) gamma ray photons.

**Neutrino**

A neutrino is an elementary particle that travels close to the speed of light, has no electric charge and only very rarely interacts with ordinary matter. They are thus extremely difficult to detect. As we will see, it is believed that neutrinos have a tiny, but non-zero mass. Its name was coined by the Italian physicist, Enrico Fermi. The first part of its name relates to it having no charge and '-ino' implies 'very small' in Italian. There are three known types (flavours) of neutrinos: electron neutrino, $v_e$, muon neutrino, $v_\mu$ and tau neutrino, $v_t$.

The first step in the proton–proton cycle is extremely slow, with a proton typically waiting $10^9$ years before carrying out this reaction! It is thus the limiting step in the chain of nuclear reactions and so determines the overall reaction rate.

It might just be worth pointing out that the slowness of this reaction is vital to our presence here on Earth. If the reaction rate was just 10 times faster, the Sun would burn up its energy supply in 1 billion years rather than the ~10 billion years that we will calculate below and there would not have been sufficient time for our human race to evolve!

**2.3.1**      *The proton–proton cycle*

The bulk of the Sun's energy comes from the main proton–proton cycle or pp I chain. This is a three-stage process:

(1)   Two protons react to give rise to a deuteron comprising one proton and one neutron. The charge of one of the protons is carried away by a positron (the antiparticle of the electron). An electron neutrino is also created.

(2)   A further proton then reacts with the deuteron to give a nucleus comprising two protons and one neutron. It is thus an isotope of helium called helium-3. A gamma ray (very high energy photon) is emitted.

(3)   Two helium-3 nuclei react to give one helium nucleus (also called an alpha particle) and two protons are emitted to take part in further reactions.

In chemical notation these steps are (Figure 2.5):

$$^1H_1 \text{ (proton)} + {}^1H_1 > {}^2H_1 \text{ (deuteron)} + e^+ \text{ (positron)} + \nu_e \text{ (electron-neutrino)}$$
$$^2H_1 + {}^1H_1 > {}^3He_2 \text{ (helium-3 nucleus)} + \gamma \text{ (gamma ray photon)}$$
$$^3He_2 + {}^3He_2 > {}^4He_2 \text{ (helium-4 nucleus)} + {}^1H_1 + {}^1H_1$$

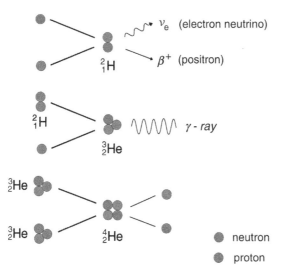

**Figure 2.5**  The three steps in the proton–proton cycle. Steps 1 and 2 are carried out twice to provide the two $^3He_2$ nuclei required for the third step.

The net result is thus:

$$4\,{}^{1}H_1 > {}^{4}He_2 + 2e^+ + 2\gamma + 2\nu_e$$

This is not quite the end of the story. The positrons given off in step 1 of the reactions will quickly meet an electron and annihilate to give two (occasionally more) further gamma ray photons. The pressure generated within the core as a result of these nuclear reactions prevents the star's collapse.

Two further reaction sequences, the ppII and ppIII chains contribute further energy; their relative reaction rates are 86% (ppI), 14% (ppII) and 0.11% (ppIII). Beryllium and lithium are formed in intermediate steps of the ppII chain and beryllium and boron in the ppIII chain. Though the ppIII chain contributes very little energy in the Sun, it does produce neutrinos of high energy which proved to be very important in the solar neutrino problem to be described later.

The photons—initially gamma rays—work their way towards the Sun's surface continuously interacting with matter. Their direction of motion following each interaction is random so they carry out what is called 'random walk' and, as a result, the energy takes the order of 100 000 years to pass through the radiative zone that surrounds the core and extends for about two-thirds of the Sun's radius. As they work their way outwards, the temperature drops and, as the photons will be in thermal equilibrium with the gas, their wavelengths will increase.

The region, about one-third of the Sun's radius in width, between the outer edge of the radiative zone and the surface, is called the convective zone. Here, the energy is carried outwards by first large and then small convective cells, as shown in Figure 2.6. As a result, the Sun's surface, called the photosphere, shows granulations – honeycomb-like variations in brightness. The granulations are brighter in the centre where the convection currents bring the energy to the surface and darker around the edges where material, cooled as it radiated energy into space, returns towards the centre.

## What percentage of the sun's mass is converted by the nuclear fusion process?

The masses of neutrons, protons and atoms are usually given in what are called atomic mass units (amu), where 1 amu is 1/12th the mass of the carbon-12 atom. (Hence, carbon-12 weighs 12 amu.) An atomic mass unit is equal to $1.66054 \times 10^{-27}$ kg. Some masses are listed below:

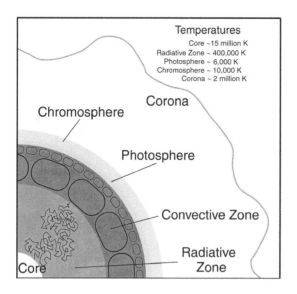

**Figure 2.6** A cross-section of the Sun showing the regions referred to in the text and their approximate temperatures.

$$\text{Electron} = 0.000549 \text{ amu}$$
$$\text{Proton} = 1.007276 \text{ amu}$$
$$\text{Neutron} = 1.008665 \text{ amu}$$
$$\text{Helium-4} = 4.0026 \text{ amu}$$

Note that a neutron is heavier than a proton and that a free neutron is unstable and will decay into a proton and an electron with a half-life of ~10.5 min giving a proton and an electron. (In other words, given a large number of neutrons, half of them will decay in 10.5 min.) This fact has great significance in the synthesis of atoms in the Big Bang and determines the relative amount of hydrogen and helium that were created. Had this half-life been significantly longer, there would be far less hydrogen in the universe and stars, like our Sun would not be able to burn hydrogen for so long.

The helium atom is made up of two protons, two neutrons and two electrons. Their individual masses add up to 4.0331 amu. However, helium-4 has a lower mass, 4.0026 amu, and is thus 0.0305 amu lighter than its components. This is the mass that has been converted into energy during the nuclear fusion process. The percentage mass loss is 0.0305/4.0026 which is thus 0.76%.

### What mass is converted to energy per second?

Energy output is $3.8 \times 10^{26}\,W$ so, as $E = mc^2$:

$$\text{Mass loss} = 3.8 \times 10^{26}\,W/(3 \times 10^8)^2\,kg\,s^{-1}$$
$$= 4.4 \times 10^9\,kg\,s^{-1}$$

However, as we have just seen, in the nuclear fusion process, only ~0.76% of the initial mass of hydrogen is converted to helium, so the amount of mass that is processed per second is $5.7 \times 10^{11}\,kg$.

### How long will the sun shine?

The Sun's mass is $2 \times 10^{30}\,kg$, so assuming that the Sun was initially composed of hydrogen and that all its mass was converted, one would get a lifetime given by:

$$\text{Lifetime} = 2 \times 10^{30}/5.7 \times 10^{11}\,s$$
$$= 3.5 \times 10^{18}\,s$$
$$1 \text{ billion years is } (365 \times 24 \times 60 \times 60) \times 1 \times 10^9\,s = 3.15 \times 10^{16}\,s$$
$$\text{So the lifetime is } 3.5 \times 10^{18}/3.15 \times 10^{16} \text{ billion years} = \text{~100 billion years.}$$

However, not all of the Sun's mass can be converted; partly as only ~75% is composed of hydrogen, but also because temperatures are only high enough (~15 million K) in the central core for nuclear fusion to take place. Theoretical models predict that only ~10% of the Sun's mass can be converted. We thus believe that our Sun will shine as a star, burning hydrogen to helium, for ~10 billion years.

### How many neutrinos are produced per second?

One proton–proton cycle transforms 0.0305 amu into energy. This corresponds to a mass given by:

$$= 0.0305 \times 1.66 \times 10^{-27}\,kg$$
$$= 5 \times 10^{-29}\,kg$$

We saw earlier that a mass of $4.4 \times 10^9\,kg$ is lost by the Sun per second, so the number of ppI cycles per second is:

$$4.4 \times 10^9/5 \times 10^{-29} = 8.8 \times 10^{37}$$

So there will be ~$10^{38}$ ppI cycles per second each producing two neutrinos, so ~$2 \times 10^{38}$ electron neutrinos will be produced per second.

It is worth noting that neutrinos carry ~2% of the energy released in the Sun, so only ~98% is radiated away from the Sun by electromagnetic radiation.

## 2.4 The solar neutrino problem

The ppI chain provides about 86% of the Sun's energy. The neutrinos given off in the ppI chain have a relatively small energy of 0.26 MeV and we have only been able to detect them recently. However, as was pointed out above, there are two further pp chains: the ppII chain produces about 14% of the solar energy whilst the ppIII chain produces only ~0.11%. The importance of the ppIII chain is that the neutrinos emitted have higher energies (from 7 up to 14 MeV) which make them easier to detect.

In the 1970s Ray Davis set up a tank filled with 100 000 gal ($4.55 \times 10^5$ l) of dry cleaning fluid, rich in chlorine nuclei, located 4900 ft (~1494 m) underground in the Homestake Mine in South Dakota (Figure 2.7). It was deep underground to prevent cosmic rays (which would be absorbed by the rock strata above) giving rise

**Figure 2.7** Ray Davis with the Homestake Mine experiment. Image: Brookhaven National Laboratory.

to false detections. Very rarely a neutrino from the ppIII chain would react with a chlorine nucleus to give a radioactive isotope of argon. Having left the tank for a month, Davis flushed the tank to extract and detect the argon atoms. On average, about 10 argon atoms were detected per month. The problem was that only about one-third of the expected number of neutrinos was detected. Remembering that these neutrinos come from relatively rare nuclear reactions in the ppIII chain, it could have been due to our lack of understanding of these reactions, but modern experiments have confirmed Davis's results. He was awarded the Nobel Prize for Physics for this work in 2002. The lack of observed neutrinos was called the solar neutrino problem.

### 2.4.1     *The solar neutrino problem is solved*

At this point we need to remember that in the standard model of particle physics there are three types of neutrino; electron, muon and tau. Those emitted by the nuclear reactions in the Sun are electron neutrinos.

At the bottom of a very deep mine at Sudbury in Canada, there is a 12 m sphere which holds 1000 t of heavy water. (Heavy water is water which has a deuteron rather than a proton as its nucleus.) Every day of operation about 10 solar neutrinos react with a deuteron, the reaction giving rise to two protons and a very high energy electron. This travels at a speed faster than the speed of light in the liquid. It thus produces a shock wave akin to that produced by a supersonic plane. This shockwave produces blue light in the form of 'Cherenkov' radiation which spreads out in a cone. The tank is surrounded by 916 photon detectors which detect the Cherenkov radiation (around 50 photons per event) and can even define the direction of the incoming neutrino so it is possible to measure the number of solar neutrinos detected. The Sudbury detector confirmed that only ~33% of the expected number of electron neutrinos was arriving from the Sun. In a follow-up experiment, 2 t of high-purity table salt (NaCl) were added to the heavy water in order to provide three times better sensitivity to the muon and tau neutrinos. It appears that the *total* number of neutrinos detected with a solar origin does agree well with that predicted for the total number of *electron neutrinos* that the Sun should produce. The only way that this could be the case is if the electron neutrinos can change (the word 'oscillate' is used) into one of the other two types en route from the Sun.

It was thought that electron neutrinos were, like photons, were mass-less. However, if the neutrino *does* have mass it can oscillate into muon or tau neutrinos. It is thought that on their way from the core of the Sun to the Earth the neutrinos will evenly distribute themselves amongst the three types – thus only one-third of the electron neutrinos emitted by the Sun will remain to be detected – exactly as observed. The neutrino problem is solved!

**2.5** ## The solar atmosphere: photosphere, chromosphere and corona

When the Sun is observed (taking great care to use the appropriate filters) it appears to have a sharp edge but there is, of course, no actual 'surface'. We are, in fact, just seeing down through the solar atmosphere to a depth where the gas becomes what is called 'optically thick'. This deepest visible layer of the atmosphere is called the photosphere (as this is where the photons that we see originate from) and is about 500 km thick (Figure 2.8). The temperature falls from ~6500 K at its base to ~4400 K at its upper region. As was derived earlier, the effective temperature of the photosphere is ~5800 K. The convective transport of energy from below gives rise to a mottling of the surface – solar granulations that are about 1000 km across. Each granulation cell last about 5–10 min as hot gas, having risen from below the surface radiates energy away, cools and sinks down again.

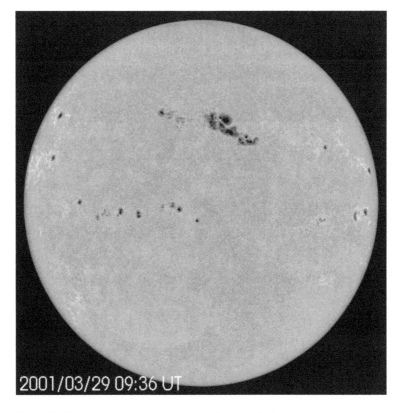

**Figure 2.8** The photosphere of the Sun showing granulations and sunspot groups. Image: SOHO space observatory, NASA, ESA.

The region, about 2000 km thick, above the photosphere, is called the chromosphere. The gas density in this region falls by a factor of about 10 000 and the temperature increases from ~4400 K at the top of the photosphere to about 25 000 K. Above this is the transition region, in which the temperature rises very rapidly over a distance of a few hundred kilometres to a temperature of ~1 million K. The transition region leads into the outer region of the Sun called the Solar Corona where temperatures reach in excess of 2 million K (Figure 2.9). Its form and extent depends strongly on the solar activity that varies through what is called the Sun Spot Cycle but, typically, extends for several solar radii into what is called the heliosphere. At the time of solar minima when activity is low it usually extends further from the Sun at its equator and the pattern of the Sun's magnetic field is often well delineated near its poles. At solar maxima, the overall shape is more uniform and has a complex structure.

The density is very low, ~$10^{14}$ times less than that at the Earth's surface, and its brightness at visible wavelengths is about a million times less than that of the photosphere. It can thus only be observed during a total eclipse of the Sun, or by using a special type of telescope, called a coronagraph, that can block out the light from the solar disc. How the Corona can reach such high temperatures is still somewhat of a mystery, but it is thought that energy might be transported into it by magnetic fields. The million degree temperatures give rise to X-ray emission that may be observed from space.

**Figure 2.9** The Solar Corona at a time close to solar minimum. Image: Fred Espernak.

### 2.5.1     *Coronium*

The spectrum of the corona – the coronal spectrum – contains emission lines. When the emission lines observed in the Solar Corona during a solar eclipse were correlated with emission lines from known elements on Earth, a green emission line was found that had no match. It was thought to relate to an unknown element and, as it had first been observed in the Solar Corona, it was provisionally named Coronium. However, in the 1930s, Walter Grotrian and Bengt Edlén, discovered that this spectral line was due to highly ionized iron – its high level of ionization being due to the extreme temperature of the Solar Corona. The resulting, very high energy, photons strip the outer electrons from atoms so giving them a positive charge. These are called ions.

## 2.6     The solar wind

The Sun is continuously 'boiling off' material in what is called the solar wind. Its existence was deduced from observations of the tails of comets. A comet usually has two tails; the most obvious is the dust tail which is composed of grains of material that are released as the Sun sublimates ice (turning it directly from a solid to a gas) from the surface of the comet nucleus and releases grains of dust embedded within it. The dust particles feel radiation pressure from the Sun and are pushed away from the Sun forming a tail which is usually curved, tending to lie along the comets' elliptical orbit. There is also an ion tail composed of ionized molecules such as cyanogen, $(CN)_2$, and carbon monoxide, CO, which gives them a bluish colour as electrons recombine. The ion tail is very straight, but cannot be explained by radiation pressure as the interaction is not sufficient. It could, however, be explained if the Sun was emitting a stream of charged particles, thought to be high energy electrons and protons, which would interact with the ions, transferring momentum to them and driving them away from the Sun.

Satellites can measure the density and velocity of the solar wind at the location of the Earth. There are about 7 000 000 ions, mostly protons and electrons but with some heavier ions, per cubic metre and these are moving at speeds of $200–700\,\mathrm{km\,s^{-1}}$. In the same way that one could calculate the total energy output of the Sun, it is possible to estimate the total mass flow away from the Sun if one assumes that the mass loss is uniform in all directions. Using a mid value for velocity of $500\,\mathrm{km\,s^{-1}}$, it appears that the Sun is losing about $3 \times 10^{-14}$ of its mass per year.

As a proton weighs ~1 amu, which is ~$1.7 \times 10^{-27}\,\mathrm{kg}$, the mass passing through $1\,\mathrm{m^2\,s^{-1}}$ is given approximately by:

$$1.7 \times 10^{-27} \times 7 \times 10^6 \times 5 \times 10^5 = 6 \times 10^{-15}\mathrm{kg}$$

As the surface area of a sphere with the radius of the Earth about the Sun is $4\pi(1.5 \times 10^{11})^2$ the approximate mass loss per year is:

$$6 \times 10^{-15} \text{kg} \times 4\pi(1.5 \times 10^{11})^2 \times 86\,400 \times 365 = \sim 5 \times 10^{16} \text{kg}$$

As the Sun has a total mass of $\sim 2 \times 10^{30}$ kg it would lose all its mass in $\sim 10^{13}$ years, but this is a thousand times longer than the predicted lifetime of $10^{10}$ years so has minimal effect on the evolution of the Sun. The interaction of the solar wind with the Earth's atmosphere will be described later.

## 2.7    The sun's magnetic field and the sunspot cycle

### 2.7.1    *Sunspots*

A photograph of the Sun's surface will usually show some darker regions on the surface. These are called sunspots and often appear in pairs or groups – a sunspot group. Each spot has a central 'umbra' surrounded by a lighter 'penumbra'. They appear dark because they are cooler than the average surface temperature. The umbra has a typical temperature 1000 K less than its surroundings. Around the outside of a sunspot group may be seen an area which is brighter than the normal surface. This is called a 'plage' – the French word for 'beach'. (French beaches often have white sand!)

The passage of sunspots across the Sun's surface has been used to measure its rotation period. It has been found that at the equator the period is ~28 days, but this increases with increasing latitude and is ~35 days near the Sun's poles, an effect called differential rotation. Sunspots are intimately linked to this differential rotation and how it affects the Sun's magnetic field.

The Sun's field must be created by the movement of charged particles and its cause almost certainly lies in the convective zone, but no really good model yet exists. Imagine that at one particular moment the Sun has a uniform bipolar field just like that shown by iron filings under the influence of a bar magnet. Looking at the Sun's face we could imagine a field line just below the photosphere directly down the centre of the Sun's disc. The field and surrounding material remain locked together in what follows. Now move forwards in time by 35 days. The field close to the poles will have made one complete rotation and be in the same position as seen from Earth but close to the equator it will have rotated an additional amount ~(35 − 28)/28 of one rotation or 1/4 × 360°, which is 90°. After three further rotations (as measured near the poles) the field near the equator will have one additional rotation. You can see that the field is being 'wound up' and becomes

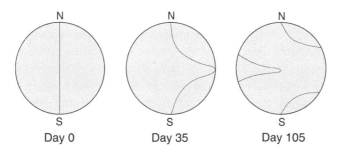

N                    N                    N

S                    S                    S
Day 0              Day 35             Day 105

**Figure 2.10**  The winding of the Sun's magnetic field due to the differential rotation of the Sun.

more intense. It gains 'buoyancy' and, in places, rises in a loop above the surface. Where it passes through the surface, the magnetic field inhibits the convective flow of energy to the surface so the localized region will be cooler than the surface in general – a sunspot appears. The energy that is inhibited from reaching the surface here will tend to reach the surface in the area surrounding a sunspot group making this region hotter and thus brighter – a plage (Figure 2.10).

It is possible to measure the polarity of the field across the Sun's surfaces and we can examine the polarity of the field associated with sunspot pairs (Figure 2.11). Consider a sunspot pair in the upper hemisphere. Assuming that the north pole is at the top then, as the field leaves the surface, it will be towards us and have

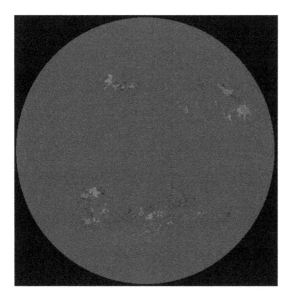

**Figure 2.11**  The polarity of sunspots as produced by a solar magnetogram. Image: Ian Morison derived from a SOHO space observatory image, ESA, NASA.

positive polarity. As the field re-enters the surface it will be away from us and have negative polarity. So the spots in a sunspot pair will show opposite polarity. In the upper hemisphere the left-hand spot would have positive polarity and the right-hand negative polarity. However, as the field reverses direction at the equator the sunspots in a pair observed in the lower hemisphere will have the opposite sense.

The twisting of the magnetic field from its initial state gradually produces an increase in the number of sunspots observed across the Sun's surface. This determines what is termed the sunspot number which reaches a peak after 3–4 years – called sunspot maximum. The field then begins to reduce in strength and the sunspot number reduces for a further 7–8 years to a point when the Sun's face can be totally devoid of spots –called sunspot minimum. (This was the case during 2007 close to sunspot minimum.)

### 2.7.2    *The sunspot cycle*

The whole process then starts over again and is called the sunspot cycle. It is often said to be an 11 year cycle, though it can vary in length somewhat and the average length of the cycle over recent decades is 10.5 years. The following cycle has, however, one important difference; the field has the opposite polarity. Hence, perhaps we should call it the 21 year solar cycle instead.

Figure 2.12 shows a plot of the sunspot numbers since 1600. The cyclical variation is very apparent. The plot shows two interesting features: a complete lack of sunspots during the late 1600s, called the Maunder minimum; and increasing solar activity during the last 50 years.

It is interesting to note that the River Thames regularly froze during the 'mini ice age' at the time of the Maunder minimum and the Earth's temperature is now rising. Could it be that there is some connection between solar activity and global temperatures? However, it should be noted that though solar activity has fallen somewhat in the last 20 years the Earth's temperature is still rising so the correlation has now been broken.

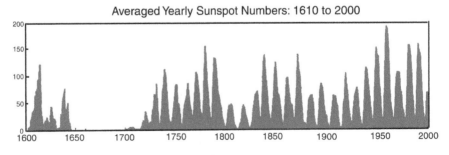

**Figure 2.12** The averaged sunspot numbers over the last 400 years.

## 2.8 Prominences, flares and the interaction of the solar wind with the earth's atmosphere

When the Sun is observed in the light emitted by excited hydrogen at a time of solar eclipse, bright columns of gas are often seen stretching up from the chromosphere into the corona. These are called prominences. They are caused by dense ionized gas that is suspended by the Sun's magnetic field but 'rains' back from the corona into the chromosphere (Figure 2.13).

Solar flares, originating in the Sun's corona, are violent explosions whose energy is believed to be derived from the 'breaking' and then 'reconnection' of the Sun's magnetic field lines and the resulting 'release' of magnetic energy. A total energy of between $10^{22}$ J and $10^{25}$ J, released over periods of minutes to hours, accelerates electrons, protons and heavier ions to relativistic speeds – that is, close to the speed of light. Flares tend to occur above the active regions around sunspots, which is where intense magnetic fields emerge from the Sun's surface into the corona, and thus tend to be more frequent at the time of solar maximum.

Flares are related to what are called coronal mass ejections, which are the ejection of material from the solar corona consisting largely of electrons and protons with small quantities of heavier elements such as helium, oxygen, and iron. The material carries with it elements of the coronal magnetic field. The streams of highly energetic particles have been observed to take as little as 15 min to reach the Earth (so travelling at about one-third the speed of light). They can pose a threat to astronauts and have, in the past, destroyed satellite sub-systems.

**Figure 2.13**  A solar prominence. Image: SOHO space observatory image, ESA, NASA.

Interference with short-wave radio communication can also occur, and the interaction of the entrained magnetic field with electricity power transmission cables can cause problems and the possible shutdown of electricity grids –as happened in the Quebec province of Canada in March 1989.

### 2.8.1     *The aurora*

One beautiful manifestation of the interaction of the solar wind with our atmosphere is coloured light displays observed in the night sky (Figure 2.14). They are most often seen within a band centred on the north and south magnetic poles and are known as the Aurora Borealis and Aurora Australis, respectively. (Aurora is the Roman goddess of the dawn, Boreas is the Greek name for the north wind, and Australis is the Latin word for south.) The Aurora Borealis is often called the northern lights as it tends to be seen as a green or reddish glow in the northern sky. It is most commonly seen around the vernal and autumnal equinoxes, though it is not known why this should be so.

Auroras are caused by the collision of charged particles with atoms high in the Earth's upper atmosphere. As the Earth's field lines open out into space above the north and south magnetic poles, charged particles may more easily reach the upper atmosphere in the regions near the magnetic poles. There they collide with atoms of gases in the atmosphere, and lift electrons into higher energy levels. The

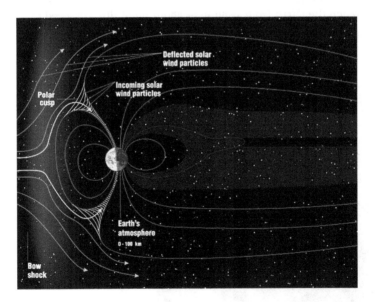

**Figure 2.14**  The path of solar wind particles towards the polar regions of the Earth. Image: Wikipedia Commons.

**Figure 2.15** An auroral display. Image: Wikipedia Commons.

electrons then cascade down to their ground states so emitting light. Most light appears to be emitted by emissions from atomic oxygen, resulting in a greenish glow at a wavelength of 557.7 nm and a dark-red glow at 630.0 nm. Amongst the many other colours that are sometimes observed, excited atomic nitrogen gives a blue colour whilst molecular nitrogen produces a purple hue. Often the auroral glow is in the form of 'curtains' that tend to be aligned in an east–west direction. Sometimes these curtains change slowly but at other times they seem to be in continuous motion. Their shape is determined by the direction of the Earth's field in the region of the observer and observations have shown that electrons from the solar wind spiral down magnetic field lines towards the Earth. The author can testify to the awesome sight that results when field lines guide electrons down to a bright auroral patch directly above the observer. Due to perspective, the converging auroral rays appear as vertical rays of light reaching upwards to what is called a 'corona' overhead (Figure 2.15).

### 2.9     Solar eclipses

Due to the tidal interactions between the Earth and the Moon, the Moon is gradually moving further away from the Earth. The fact that it now has an angular size which is usually just greater than that of our Sun gives rise to what is probably the most spectacular of celestial events – a total solar eclipse.

To give a total eclipse, the Moon must lie in the plane of the Solar System (the plane containing the Earth and Sun), be at new moon and have an angular size greater than that of the Sun. The Moon's orbit is inclined at 5° to the plane of the Solar System, so it will usually be above or below the plane at the time of new moon so we do not get solar eclipses every month. Typically, there will be a solar eclipse visible somewhere on the Earth every 18 months. However, an eclipse is not necessarily total. Both the Earth and the Moon have elliptical orbits. If the Moon is furthest from the Earth (at what is called apogee, so making the Moon's angular size smaller) and the Earth is closest to the Sun (at perihelion, so making the Sun's angular size bigger) then the Moon cannot cover the Sun completely and we get what is called an annular eclipse when a ring of the Sun is still visible around the edge of the Moon. This is often called a 'ring of fire' and is particularly spectacular when seen close to dawn or dusk when the Sun's light is reddened.

During a total eclipse the Moon's shadow will first touch the Earth's surface at the start of what is termed the 'eclipse track' (Figure 2.16). The width and length of the track will depend on positions of the Moon and Earth in their orbits. Close to the Earth's equator, the Moon is nearer to the Earth's surface than at the poles so its angular size is greater and the eclipse track is wider, so it is here that the longest period of totality can be observed. The longest possible total eclipse is approximately 7 min 30 s long. This would be observed at the equator when the Earth is at aphelion (farthest from the Sun) and the Moon is at perigee (closest to the Earth). The longest eclipse that may be observed in the next 3000 years is on July 16, 2186 and this will have a duration of 7 min 29 s. The longest in recent times was observed by many from Hawaii on July 11, 1991 when totality lasted 6 min 53 s.

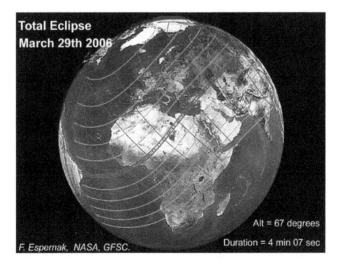

**Figure 2.16** The eclipse track during the solar eclipse of March 2006. Image: Fred Espernak, NASA.

It is not a coincidence that the dates within the year are so close. The Earth is at aphelion on July 4 so the longest periods of totality will occur close to that date.

The ancients observed that sets of eclipses, solar and lunar, recurred every 18 years 11 days and 8 h – a time period called the Saros. As a result of the one-third of a day in this period, each set of eclipses is observed one-third of the way around the globe at successive Saros periods. Solar eclipses will thus be seen in the same region of the Earth after three Saros periods (54 years and 34 days).

## 2.9.1    *Two significant solar eclipses*

The eclipses of 1919 and 1922 played a significant role in the history of science. As was described in Chapter 1, Einstein predicted that the Sun's mass would distort the space–time around it and thus light would travel along curved lines when passing close by. One observable effect that would allow his radical theory to be tested was that this should cause the positions of stars, when seen in close proximity to the Sun, to be shifted from their true positions. (It should be noted that Newton also predicted, for different reasons, that light waves should be bent when passing the Sun, but the effect due to the distortion of space–time in Einstein's theory is twice that predicted by Newton's theory.)

In essence the plan was simple. Prior to a solar eclipse, take images of the sky where the Sun would be during totality. Then take the same images during totality – the only time when stars can be seen close to the Sun's position – and compare the positions of the stars. Close to the limb Einstein's theory predicts a shift of just 1.75 arcsec (while Newton's suggested 0.8 arcsec).

Sir Arthur Eddington led the British eclipse expedition to the Atlantic Island of Principe in order to observe the total eclipse of the Sun in May 1919, whilst a second set of observations were made from Sobral in Brazil (Figure 2.17). The telescopes used thus had to be portable and this limited their accuracy. The control images obviously had to be taken at night when it would have been colder than during the day time. Even disregarding these problems, the experiment was not easy. The anticipated deflection of stars nearby the Sun (a maximum of 1.75 arcsec) has to be compared with the typical size of a stellar image as observed from the ground (due to atmospheric turbulence) of 1–2 arcsec. The data from the observations were not quite as conclusive as was implied at the time. The telescope at Principe was used to take 16 plates, but partial cloud reduced their quality. Two usable plates from the telescope on Principe, though of a poor quality, suggested a mean of 1.62 arcsec. Two telescopes were used at Sobral where conditions were superb; sadly, however, the focus of the main instrument shifted, probably due to temperature changes, and the stellar images were not clear. They were thus difficult to measure and produced a result of ~0.93 arcsec. A smaller 10 cm instrument did, however, produce eight clear photographic plates and these showed a mean

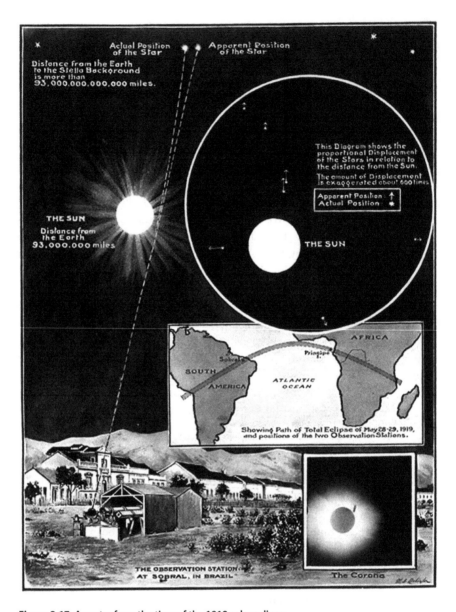

**Figure 2.17** A poster from the time of the 1919 solar eclipse.

deviation of $1.98 +/- 0.12$ arcsec. If all the data had been included, the results would have been inconclusive, but Eddington, with little justification, discounted the results obtained from the larger Sobral telescope and gave extra weight to the results from Principe (which he had personally recorded). On November 6 that year the Astronomer Royal and the President of the Royal Society declared the evidence was decisively in favour of Einstein's theory. However, there were many scientists who, at the time, felt there were good reasons to doubt whether the observations had been able to accurately test the theory.

A more positive test of the theory came from observations made by William Campbell's team from the Lick Observatory who observed the 1922 eclipse from Australia. They determined a stellar displacement of 1.72 +/− 0.11 arcsec. Campbell had believed that Einstein's theories were wrong, but when his experiment proved exactly the opposite, he immediately admitted his error and thereafter supported relativity. (One tends to believe an experiment when the results do not agree with the expectations of the observer!)

More recent results determining the deflection of light and radio waves have come from the Hipparcos satellite and very precise positional measurements made by arrays of radio telescopes across the globe. The Hipparcos satellite could not observe star positions near to the Sun but, even at 90° away from the Sun, the deflection is 4 milliarcsec – within its measurement capability when averaged over many stars. Observations of thousands of stars have given results that agree with Einstein's predictions to ~0.3%. Radio measurements comprising over 2 000 000 observations are in agreement with his predictions to 0.02%.

## 2.9.2    The Shapiro delay

More recent verifications of Einstein's theories relating to the curvature of space–time due to the Sun's mass have been made with planetary radar observations of the planet Mars. In the 1960s, Irwin A. Shapiro realised that there was another, and potentially far more accurate, way of testing Einstein's theory. Shapiro was a pioneer of radar astronomy and realised that the time that a radar pulse would take to travel to and from a planet would be affected if the pulse passed close to the Sun. Figure 2.18a first shows the direct path that a radar pulse would take to and from Mars if we could imagine that the Sun was not present and that, as a consequence, the radar pulse travelled in a direct straight line. Figure 2.18b then shows that, due to the curvature of space by the Sun, a radar pulse sent along this precise path would not reach Mars, but curve away to the left. The pulse that *would* reach Mars, shown in Figure 2.18c, has to take a path slightly to the right of its true position so the curvature of space near the Sun would deflect it towards Mars. The echo would follow exactly the same path in reverse. As the pulse has had to follow a longer route to Mars and back it will obviously take longer than if the Sun was not present. The radar pulse will thus be delayed. The Shapiro delay, as it is called, can reach up to 200 μs and has provided an excellent test of Einstein's theory.

Further tests, of even higher accuracy, using the Shapiro delay have been made by monitoring the signals from spacecraft as the path of the signals passed close

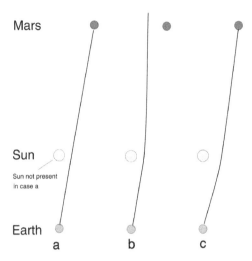

**Figure 2.18** The Shapiro delay.

to the Sun. In 1979, the Shapiro delay was measured to an accuracy of one part in a thousand using observations of signals transmitted to and from the Viking spacecraft on Mars. More recently, observations made by Italian scientists using data from NASA's Cassini spacecraft, whilst en route to Saturn in 2002, confirmed Einstein's General Theory of Relativity with a precision 50 times greater than previous measurements. At that time the spacecraft and Earth were on opposite sides of the Sun separated by a distance of more than 1 billion km (approximately 621 million miles). They precisely measured the change in the round-trip travel time of the radio signal as it travelled close to the Sun. A signal was transmitted from the Deep Space Network station in Goldstone, California which travelled to the spacecraft on the far side of the Sun and there triggered a transmission which returned back to Goldstone. The use of two frequencies to make simultaneous measurements enabled the effects of the solar atmosphere on the signal to be eliminated so giving a very precise round-trip travel time. The Cassini experiment confirmed Einstein's theory to an accuracy of 20 parts per million.

## 2.10    Questions

1.  The peak of the spectrum of a star (assumed to be a black body radiator) is observed to be at a wavelength of $0.7 \times 10^{-6}$ m. Estimate the star's surface temperature.
2.  A star has a luminosity of $8 \times 10^{26}$ W and a diameter of $8 \times 10^{8}$ m. Use the Stefan–Boltzmann Law to estimate its surface temperature.

3.  A star has twice the diameter and twice the surface temperature of our Sun. How will its luminosity compare with that of our Sun?

4.  A planet lies at a distance of $3 \times 10^{11}$ m from the central star of its solar system. The solar constant is measured to be $280\,\text{W}\,\text{m}^{-2}$. What is the total energy output of the star? The angular diameter of the star, as measured from the planet, is 22 arcmin. Estimate the surface temperature of the star.

5.  An eclipse of the Sun is to occur on the day of the vernal equinox and when the Earth is at a distance of 150 million km from the Sun. The eclipse track passes over the equator at 12:00 Local Solar Time. Show that, if the Moon is at its furthest point from the Earth (at apogee) that day, the eclipse as observed from the equator will be an annular, not a total, eclipse.

    Show also that if the Moon were to be closest to the Earth that day (at perigee) there *would* be a total eclipse.

    (The equatorial radius of the Earth is 6378 km. The semi-major axis of the Moon's orbit is 384 401 km and the eccentricity of its orbit is 0.056. Its diameter is 3474.8 km. The diameter of the Sun is 139 million km.)

6.  Following the summer exams, two astrophysics students had been sunbathing on a flat beach in Sweden (latitude: $+62°$) at 12:00 Local Solar Time on June 21. In the evening, the students decided to estimate the number of neutrinos per second passing through their bodies that morning. Their lecture notes told them that the Earth's axis is tilted at $23.5°$ to the plane of the ecliptic, the solar constant was $\sim 1300\,\text{W}\,\text{m}^{-2}$, that one proton–proton cycle converted $4.6 \times 10^{-29}$ kg of mass into energy and they knew that the velocity of light is $3 \times 10^8\,\text{m}\,\text{s}^{-1}$. The students realised that it was not necessary to know how far the Earth was from the Sun and quickly came up with a result. What might the result have been, assuming that the cross-sectional area of a prone student was $1\,\text{m}^2$?

    (Note: you can get a pretty good result ignoring where and when they were, but some of you might like to give a more accurate result. If the Earth's equator was in the plane of the Solar System, the Sun would remain at declination zero. Given a tilt of the Earth's rotation axis of $\theta°$, the Sun's declination will vary between $+\theta°$ and $-\theta°$ over the year.)

## Chapter 3

# Our Solar System 2 – The Planets

The first part of this chapter will look at our own Solar System as an example of Solar Systems in general (Figure 3.1). It will discuss aspects of planetary orbits, how the Sun's radiation and the properties of the planets determine their surface temperatures and how atmospheres form and change during the lifetime of a planet.

The second part will give a survey of the individual planets of our Solar System, highlighting interesting aspects related to their properties, discovery and satellites. Such a survey could (and often does) have a complete book devoted to it, and the student is encouraged to use such books or the Web to learn more.

## 3.1 What is a planet?

This question was not seriously considered until a meeting of the International Astronomical Union (IAU) in August 2006. There had never been a formal definition of what should, or should not, be a planet and for some time the minor planet Ceres, which is the largest body in the asteroid belt between Mars and Jupiter, had also been classed as a planet.

In 2005, the discovery of a body (initially called 2003 UB$_{313}$) was announced. It is slightly larger than Pluto and was at a distance of 96.7 AU from the Sun (three times the distance of Pluto). This required a decision as to whether it should become the 10th planet of the Solar System or whether, instead, Pluto should be demoted. Pluto is considerably smaller than it was thought to be when first discovered and has a highly elliptical orbit inclined at a large angle to the plane of the Solar System. If it had been discovered in recent times, it is highly unlikely that it would have been given the status of a planet. For these reasons, the Hayden Planetarium in New York had already, with some controversy, omitted Pluto from its planet gallery.

*Introduction to Astronomy and Cosmology*   Ian Morison
© 2008 John Wiley & Sons, Ltd

**Figure 3.1** The eight planets of our Solar System. Image: Wikipedia Commons, NASA.

The definition that was agreed upon in August 2006 had three parts:

(1) A planet orbits the Sun.
(2) A planet has enough mass so that gravity can overcome the strength of the body and so becomes approximately round. It is said to be in hydro-static equilibrium.
(3) A planet has 'cleared' its orbit – that is, it is the only body of its size in the region of the Solar System at that distance from the Sun.

The first two parts are fairly obvious, but the third is less so. Essentially it means that there must not be other comparable sized objects orbiting the Sun at similar distances.

It is part (3) that demotes Pluto.

The IAU also produced a definition of what will become known as dwarf planets. These satisfy parts (1) and (2) of the definition of a planet, but not part (3).

In addition they must not be the satellite of another body. As a result Pluto is now classed as a dwarf planet along with 2003 UB$_{313}$, which has now been given the name Eris. The minor planet Ceres also satisfies the definition of a dwarf planet so currently we have eight planets and three dwarf planets in the Solar System. It is likely that, over time, the number of dwarf planets will increase as further large objects are discovered in the region beyond Neptune.

## Planetary orbits

Figure 1.13 shows the terms that are given to the orbital properties of the planets. They orbit the Sun in elliptical orbits with the Sun at one focus. A key parameter is the semi-major axis, $a$, which is half the major axis of the ellipse. For a circular orbit, $a$ will be the radius of the orbit. Due to the eccentricity, $e$, of their orbits their distance from the Sun varies from a minimum distance (at perihelion) of $a(1-e)$ to a maximum distance (at apehelion) of $a(1 + e)$.

In our Solar System Venus, whose orbit has an eccentricity of 0.007, is in a virtually circular orbit. Neptune and the Earth have near circular orbits with eccentricities of 0.01 and 0.17, respectively. Mercury and the dwarf planets Pluto and Eris have the most eccentric orbits with eccentricities of 0.205, 0.249 and 0.441, respectively. It is worth noting that, for part of its orbit, Pluto can come closer to the Sun than Neptune.

An interesting consequence of the eccentricity of the orbit of Mars is that when Mars comes closest to us (every 2 years and 2 months) its distance from us can vary significantly. As a result, its angular size at closest approach will vary significantly and so determine the amount of surface detail that we can see from Earth. Mars is closest to us within a few days of opposition – that is, when it is on the opposite side of the sky to the Sun – and will thus be seen approximately south at midnight. Mars will be seen with the smallest angular size at opposition when Mars is furthest from the Sun (at aphelion) and the Earth is closest from the Sun (at perihelion) as shown in Figure 3.2a. Figure 3.2b shows the inverse situation when Mars will be seen with the largest angular diameter during opposition.

The Earth is at aphelion, furthest from the Sun, on July 4 each year, so the very closest approaches of Mars will occur in the summer months. The closest approach for nearly 60 000 years occurred on August 27, 2003, when Mars was 55 758 006 km from Earth and had an angular diameter of just over 25 arcsec (Figure 3.3). In contrast, if Mars is at aphelion and the Earth at perihelion at the time of closest approach, as shown in Figure 3.2b, then the angular size is just less than 14 arcsec – a very significant difference!

The angular sizes observed at opposition are currently reducing and reach a minimum of 13.89 arcsec on March 3, 2012. They will then increase again until,

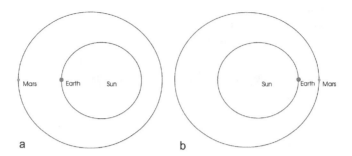

**Figure 3.2** The situations when Mars is seen with the smallest (a) and largest (b) angular sizes when at opposition.

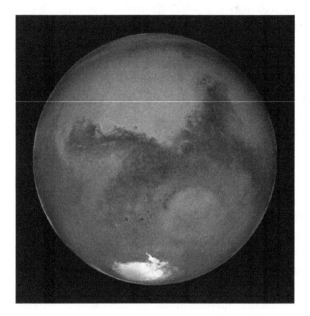

**Figure 3.3** Mars as observed by the Hubble Space Telescope at the time of its closest approach for 60 000 years. Image: J. Bell (Cornell U.), M. Wolff (SSI) *et al.*, STScI, NASA.

on July 27, 2018, the angular diameter will be 24.31 arcsec – only just less than the absolute maximum.

### 3.2.1    *Orbital inclination*

The orbital inclination of a planet's orbit is the angle at which the orbital plane of a planet is inclined to the plane of the Solar System. The plane of the Solar System is defined to include the Earth, so that the orbital inclination of the Earth's orbit is zero. The inclination angles tend to be small except in the case of Mercury, at 7°, and the dwarf planets, Pluto and Eris, at 17 and 44.2°, respectively.

## 3.3  Planetary properties

### 3.3.1  *Planetary masses*

It is possible to find the mass of a planet if:

(1)  It has one or more natural satellites in orbit around it, as in the case for the planets Earth, Mars, Jupiter, Saturn, Uranus and Neptune and the dwarf planets Pluto (Charon) and Eris (Dysnomia).
(2)  It has acquired an artificial satellite as in the case of the Magellan space-craft in orbit about Venus.
(3)  It has been passed by an artificial satellite as was the case when Mariner 10 flew by Mercury.

It has thus been possible to calculate the mass of all the Solar System planets and two of its dwarf planets. In September 2007, the Dawn Mission spacecraft was launched to visit both Ceres and a second body within the asteroid belt, Vesta, so we will then have masses for all currently known planets and dwarf planets.

The method is essentially that used in Chapter 2 to calculate the mass of the Sun using the period and semi-major axis of the Earth's orbit. Let us take Mars as an example.

### Calculating the mass of mars

Mars has a satellite, Phobos, which orbits Mars with a period of 7 h 39.2 min (27 552 s) in an almost circular orbit having a semi-major axis of 9377.2 km or $9.3772 \times 10^6$ m.

The gravitational force between Mars and Phobos is given by $MmG/a^2$, where $M$ is the mass of Mars, $m$ is the mass of Phobos, $a$ is the semi-major axis of Phobos's orbit and $G$ the universal constant of gravitation.

This force must equal that resulting from the centripetal acceleration, $\omega^2$, or $v^2/a$ on Phobos of mass $m$:

$$MmG/a^2 = mv^2/a$$

Giving:

$$M = v^2 a/G$$

But $v = 2\pi\, a/P$, where $P$ is the period of Phobos, so substituting:

$$M = 4\pi^2 a^3/GP^2$$

with units of kilograms, seconds and metres.

So

$$M = 4 \times (3.14159)^2 \times (9.3772 \times 10^6)^3/[6.67 \times 10^{-11} \times (2.7552 \times 10^4)^2]\,kg$$
$$= 6.43 \times 10^{23}\,kg$$

Having completed this calculation as this text was being written, the author was very gratified to find that the accepted value for the mass of Mars is $6.42 \times 10^{23}$ kg. The calculation used the slight approximation that the orbit of Phobos was circular so perfect accuracy should not have been expected.

### 3.3.2      *Planetary densities*

From the angular size of a planet and its distance one can calculate the diameter of a planet and hence its volume so, given its mass, one can thus calculate its density. It is interesting to use Saturn as an example. Saturn is not a sphere, but an oblate spheroid having a greater equatorial than polar radius. We will use an 'average' value of ~59 000 km for our calculation. Its volume will thus be $4/3 \times \pi r^3$:

$$V = 4/3 \times 3.14159 \times (5.9 \times 10^7)^3\,m^3$$
$$= 7.76 \times 10^{23}\,m^3$$

Given Saturn's mass of $5.7 \times 10^{26}$ kg, this gives a density of ~$662\,kg\,m^{-3}$, which is a little less than the accepted value of $687\,kg\,m^{-3}$. You will note that this is less than that of water ($1000\,kg\,m^{-3}$ at $4°C$)!

### 3.3.3      *Rotation periods*

For some planets, such as Mars, Jupiter and Saturn, one can observe the rotation of a marking on the surface or in the atmosphere – such as the 'red spot' which lies in the atmosphere of Jupiter.

The surface of Mercury is very indistinct as seen from Earth and Venus is cloud covered! In these two cases, planetary radars have been able to measure the rotation periods. Imagine the point at which Venus lies exactly between the Earth and the Sun and a continuous radio frequency is reflected from Venus when a radar transmitter lies on the line between the centre of the Earth, through the centre of Venus to the centre of the Sun. At this time (and this time only) the motion of Venus will be across the line of sight and there will be no Doppler shift in the returned echo. (To be totally accurate, there will be a *very* small Doppler shift called the transverse Doppler shift due to a prediction of special relativity.)

The centre frequency of the returned echo will be at precisely the same frequency as that transmitted and, if Venus is *not* rotating, all the returned energy will be at this frequency. Suppose, however, Venus *is* rotating. One limb will be coming towards us relative to the centre, whilst the other will be moving away from us. The echoes from the limbs will thus be Doppler shifted above and below the centre frequency so that the returned echo is 'broadened' in frequency. The greater the broadening, the greater the rotation rate of the planet.

Radar observations made in the 1960s showed that Venus had a very slow rotation rate, taking 243.01 days to rotate once round its axis – 18.3 days longer that it takes to orbit the Sun! Even more surprising, it rotates in the opposite direction to that expected. Looking down from above the Solar System all the planets rotate in an anticlockwise direction. The spin of most planets is also in an anticlockwise direction, but Venus (along with Uranus and Pluto) rotates in a clockwise direction, in the opposite sense to its orbital motion, and the rotation is said to be retrograde.

**3.3.4**    *Planetary temperatures*

There are three ways that we can measure or estimate the surface temperature of a planet:

(1)  In the case of Venus and Mars, spacecraft on the surface have made direct measurements.
(2)  The temperature of Mercury was estimated from the intensity of its radio emission assuming it to act as a black body. In a similar way, the temperatures of the outer planets can be estimated from their infrared emission.
(3)  We can calculate a nominal temperature on the assumption that a planet acts like a black body and will radiate away the energy that it receives from the Sun. (There must be an equilibrium between the energy absorbed from the Sun and that emitted by a planet.)

We will consider this last method in more detail and carry out a calculation to derive the surface temperature of the Earth.

We know that, above the atmosphere, 1368 W of solar energy (the solar constant) fall on the Earth per square metre. Figure 3.4 shows that the Earth will

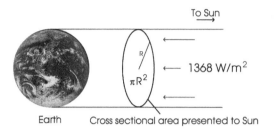

**Figure 3.4** The Earth's effective area for absorption of solar energy.

intercept this radiation over an area given by the Earth's cross-section. If $SC$ is the value of the solar constant, then the total energy that will fall on the Earth is given by:

$$\pi R^2 SC$$

If we assume that the planet acts as a black body, then the energy emitted by it is given by the Stefan–Boltzmann Law and equals:

$$4\pi R^2 \sigma T^4$$

In equilibrium, these energies must be equal:

$$\pi R^2 SC = 4\pi R^2 \sigma T^4$$

So

$$SC = 4\sigma T^4$$
$$T = (SC/4\sigma)^{1/4}$$
$$= [1368/(4\times 5.7\times 10^{-8})]^{1/4}$$
$$= 278\,K$$

This is not far off the actual average surface temperature, but should it be?

The Earth is, on average, about 50% cloud covered and absorbs only ~77% of the incident solar radiation from the Sun. Taking this into account by reducing the incident energy by 0.77, $T_{Earth}$ would only be ~260 K. However, the Greenhouse Effect produced by the carbon dioxide, methane and water vapour in the atmosphere prevents the Earth radiating away as much energy as would a perfect black body, so increasing the Earth's temperature. The two effects roughly cancel out giving us an average temperature for $T_{Earth}$ of ~288 K. It is worth pointing out that without the greenhouse gases in our atmosphere our planet would be uninhabitable.

Greenhouse gases absorb infrared radiation emitted by the Earth and then re-emit it in random directions – so much of the infrared radiation will thus be directed back at the Earth!

### 3.3.5     *Global warming*

The major constituents of the atmosphere, nitrogen, $N_2$, and oxygen, $O_2$, are not greenhouse gases. This is because diatomic molecules such as these neither absorb nor emit infrared radiation. Carbon dioxide is the main greenhouse gas in the atmosphere. Over aeons of time its percentage in the atmosphere has remained stable but, unfortunately, the burning of fossil fuels (which have stored carbon within them) is rapidly increasing the amount of carbon dioxide in the atmosphere and this is almost certainly a major contribution to the fact that our Earth's temperature is increasing – termed global warming or climate change.

Water vapour is a naturally occurring greenhouse gas and actually accounts for the largest percentage of the greenhouse effect, somewhere between 36% and 66%. The amount of water vapour in the air from locality to locality is very variable but overall, human activity does not directly affect water vapour concentrations (except near irrigated fields for example) and its effects on the Earth's climate are remaining stable.

However, the amounts of two further greenhouse gases are now also increasing:

(1)     Methane is 20 times more efficient at retaining heat than carbon dioxide and we are adding up to 500 million t of methane into the atmosphere per year from livestock, coal mining, drilling for oil and natural gas, rice cultivation, and garbage decaying in landfills.

(2)     Each year, 7–13 million t of nitrous oxide is added to the atmosphere from the use of nitrogen based fertilizers, the disposing of human and animal waste in sewage treatment plants and automobile exhausts.

An increase in the Earth's average temperature of more than 2 degrees could begin to have very harmful consequences for the human race, which explains why the problem is being treated so seriously.

### 3.3.6     *Albedo*

As the example of the Earth has shown, the actual temperature of a planet is affected by how much of the Sun's incident energy is reflected back into space – called the albedo of a planet – and the effects of greenhouse gases, if any.

The Earth has an albedo of ~0.37, meaning that it reflects ~37% of the Sun's energy and so will absorb 63%. Venus has an albedo of ~0.7 (published values vary from 0.65 to 0.84) so that it only absorbs 30% of the incident solar energy, but its carbon dioxide atmosphere is so thick that its surface temperature is raised significantly. Mars has an albedo of 0.15 so absorbs much of the incident solar energy but its thin carbon dioxide atmosphere (about 1/100th that of the Earth) is unable to trap much heat so it is now too cold for carbon/water based life forms to survive on the surface. However, in the past, when giant volcanoes were emitting vast amounts of gas into the atmosphere (including water vapour, carbon dioxide and methane) its temperature would have been significantly higher and life could, perhaps, have arisen there.

## 3.4    Planetary atmospheres

This section will discuss the formation and evolution of planetary atmospheres. In every case the planet's original atmosphere will have come from the solar nebula out of which the Sun and planets formed. Its composition will thus have been similar to that of the Sun so that it will have been largely composed of the light elements hydrogen and helium. This was the only source of the atmospheres of the outer planets but may not have contributed very much to the terrestrial planets as, by the time they had formed, the solar wind of the young star would have ejected much of the solar nebula outwards beyond the inner planets. In addition, as will be seen below, an atmosphere of light gases could not have been kept by these relatively small planets due to their relatively high surface temperatures and low gravity.

Consider an atmosphere made up of a number of different gases, some of light molecules, such as hydrogen and helium, and some of heavier molecules, such as carbon dioxide, ammonia and methane. The law of Equipartition of Energy states that all species of molecules in the atmosphere will have roughly equal kinetic energies ($\frac{1}{2}mv^2$). This means that for a given temperature, lighter molecules with smaller masses, will have higher velocities than heavier molecules. The average kinetic energies of the gas molecules will depend on the temperature of the atmosphere so, in hotter atmospheres, the molecules will be moving faster.

The average kinetic energy of motion of a molecule is related to the absolute temperature, $T$, by:

$$1/2mv^2 = 3/2kT$$

where $k$ is Boltzmann's constant which is equal to $1.38 \times 10^{-23} \, \mathrm{m^2 \, kg \, s^{-2} \, K^{-1}}$.

Giving:

$$v = (3kT/m)^{1/2}$$

For a given temperature (and hence kinetic energy), the velocity of a given molecule will be inversely proportional to the square root of its molecular mass, so molecules of hydrogen (molecular mass 2) will move on average four times faster than those of oxygen (molecular mass 32). If a molecule in the upper part of an atmosphere happens to be moving upwards at a sufficiently high velocity, then it could exceed the escape velocity of the planet and so escape into space. The escape velocity depends on the mass of the planet so it should be apparent that hot, light planets might well lose all the lighter molecules that they might have once have had in their atmospheres whilst cooler, more massive planets will be able to hold on to even the lightest molecules within their atmospheres.

Let us calculate the average velocity for nitrogen molecules which have a mass of $4.68 \times 10^{-26}$ kg:

$$v = [(3 \times 1.38 \times 10^{-23} \times 300)/4.68 \times 10^{-26}]^{1/2}$$
$$= 0.515 \, km \, s^{-1}.$$

So, for nitrogen (molecular mass 28) and oxygen (molecular mass 32) in the Earth's atmosphere at a temperature of ~300 K, the typical molecular speeds are $0.52$ and $0.48 \, km \, s^{-1}$, respectively. This is far smaller than the escape velocity of the Earth which is $11.2 \, km \, s^{-1}$ so we would not expect these gases to escape from our atmosphere. In fact, it is not quite as simple as that. Due to collisions between them, molecules do not all move at the same speed; some are faster and some slower than the average. The relative numbers of molecules at speeds around the average is given by the Maxwell–Boltzmann distribution. A very small fraction of the molecules in a gas have speeds considerably greater than average, with one molecule in 2 million moving faster than three times the average speed, and one in $10^{16}$ exceeding the average by more than a factor of 5. Consequently, a very few molecules may be moving fast enough to escape, even when the average molecular speed is much less than the escape velocity. Calculations show that if the escape velocity of a planet exceeds the average speed of a given type of molecule by a factor of 6 or more, then these molecules will not have escaped in significant amounts during the lifetime of the Solar System.

In the Earth's atmosphere, the mean molecular speeds of oxygen and nitrogen are well below one-sixth of the escape speed. Now consider the Moon: its escape velocity is $2.4 \, km \, s^{-1}$ and, assuming that any atmosphere it might have had would have been at the same temperature as our own, the mean molecular speeds of nitrogen and oxygen would be only about five times less than the Moon's escape velocity so it is not surprising that it has no atmosphere! If Mercury had an atmosphere it would have a temperature of ~700 K, giving nitrogen or oxygen an average molecular speed of about $0.8 \, km \, s^{-1}$, significantly more than one-sixth of Mercury's escape velocity of $4.2 \, km \, s^{-1}$. There has thus been ample time for these molecules to escape.

These arguments allow us to see why our own atmosphere contains very little hydrogen. Hydrogen molecules move, on average, at about $2 \, km \, s^{-1}$, which is just more than one-sixth of the Earth's escape velocity. Hydrogen will thus have been able to escape and now makes up only 0.000055% of the atmosphere! In contrast, consider Jupiter: its escape velocity is $60 \, km \, s^{-1}$ and it has a surface temperature of only 100 K. In the Jovian atmosphere, the speed of the hydrogen molecules is only about $1 \, km \, s^{-1}$, 60 times less than the escape velocity, and so Jupiter has been able to keep the hydrogen as the largest constituent in its atmosphere.

### Summary

- Mercury, the Moon and all satellites except for Titan and Triton have effectively no atmospheres, though Mercury has an extremely thin 'transient' atmosphere of hydrogen and helium temporarily captured from the solar wind.
- The other terrestrial planets cannot hold on to hydrogen or helium, so will have lost all the initial atmospheres derived from the solar nebula.
- The outer planets are both massive and cold and so have been able to keep all of the light gases acquired from the solar nebula. Though similar in mass to the Moon, Titan and Triton are sufficiently cold to have kept atmospheres, largely made up of nitrogen.
- The dwarf planets Pluto and Eris are so cold that any nitrogen or other gases would be frozen and form part of the surface.

### 3.4.1    *Secondary atmospheres*

Later in their lives, the planets Venus, Earth and Mars gained further atmospheres which were the result of out-gassing from volcanoes. It is thought that only 1% of the current Earth's atmosphere remains from its primeval atmosphere. Volcanic eruptions produce varying amounts of gases which arise from the melting of the planet's crust at great depth. All eruptions differ but, in general, release gases such as water vapour, carbon dioxide, sulphur dioxide, hydrogen sulphide, ammonia, nitrogen and nitrous oxide. It is thought that ultraviolet light falling on water vapour in the upper atmospheres of Venus and Mars would have split it into hydrogen and hydroxyl (OH). The hydrogen molecules would then have escaped so removing water vapour from their atmospheres.

**3.4.2**    *The evolution of the earth's atmosphere*

We have seen how the primeval atmosphere of the Earth, largely made up of hydrogen and helium would have been lost and replaced by a secondary atmosphere which was the result of volcanic out-gassing. It was primarily made up of carbon dioxide and water vapour, with some nitrogen but virtually no oxygen and contained perhaps 100 times as much gas as at present. As the Earth cooled, much of the carbon dioxide dissolved into the oceans and precipitated out as carbonates.

A major change began some 3.3 billion years ago when the first oxygen producing bacteria arose on Earth and which, in the following billion years, gave us much of the oxygen in our atmosphere. Oxygen and bacteria could then react with the ammonia released by out-gassing to form additional nitrogen. More nitrogen was formed by the action of ultraviolet radiation on ammonia in a process call photolysis.

As the vegetation increased, the level of oxygen in the atmosphere increased significantly and an ozone layer appeared. This gave protection to emerging life forms from ultraviolet light and enabled them to exist on land as well as in the oceans. By about 200 million years ago about 35% of the atmosphere was oxygen. The remainder of the atmosphere was largely nitrogen as, alone of all the gases present in the secondary atmosphere, it does not readily dissolve in water.

Volcanic activity recycles and replenishes the molecules of the atmosphere and has, in particular, recycled the greenhouse gas carbon dioxide – necessary for the Earth's surface temperature to have remained sufficiently warm for the existence of life. The carbonates, formed as carbon dioxide dissolves in the oceans, and shells of calcium carbonate produced by marine life, fall to the ocean beds. Thus, over time, one might expect the amount of carbon dioxide in the atmosphere to reduce, tending to make the Earth colder. However, movements of the oceanic plates of the Earth's crust bring them up against the continental plates. As the oceanic plates are denser, they pass under the continental plates – a process known as subduction – and volcanic activity releases the carbon dioxide back into the atmosphere.

**3.5**    **The planets of the solar system**

Our knowledge of the Solar System has increased dramatically in the last 50 years by the use of spacecraft that have flown by, orbited and even landed on the planets giving us high resolution images and detailed planetary data. This section will summarize some of the most interesting aspects of the individual planets of our Solar System but does not, in any way, pretend to be exhaustive. The information included is that which has been found of most interest by the author and he can only hope that you will find it interesting too.

### 3.5.1     *Mercury*

| Mass | $3.3 \times 10^{23}$ kg | 0.055 that of Earth |
|---|---|---|
| Mean radius | 2439.7 km | 0.383 that of Earth |
| Escape velocity | 4.25 km s$^{-1}$ | |
| Rotation period | 58.646 days | |
| Semi-major axis of orbit | 57 909 068 km | 0.387 AU |
| Average orbital speed | 47.87 km | |
| Eccentricity | 0.2053 | |
| Period | 87.97 days | |
| Orbital inclination | 7° | |

Due to Mercury's close proximity to the Sun it may only be observed for relatively short periods at a time and, seen from the Earth, its surface markings are not very distinct. Both these factors contributed to an initially erroneous value for its rotation period which was thought to be the same as its orbital period. Mercury would then have been tidally locked to the Sun with one face continuously towards, and one face away, from the Sun. Radio observations indicated that Mercury's 'dark' side was considerably warmer than would have been expected should it have always faced away from the Sun and radar studies have since shown that Mercury rotates about its axis with a period two-thirds that of its orbital period.

Mercury has only ever been visited by two spacecraft: Mariner 10, which made three fly-bys in the mid 1970s; and the Messenger spacecraft, which flew by early in 2008 (Figure 3.5). Mercury looks very like the highland regions of the Moon.

In 1991, Mercury was observed by radar using the 70-m Goldstone antenna with a half-million watt transmitter. The radar reflections were received by the Very Large Array in New Mexico to provide high resolution radar images. To great surprise, a very strong reflection was received from Mercury's North Polar region which strongly resembled the strong radar echoes seen from the ice-rich polar caps of Mars. At very low temperatures water ice is a very effective radar reflector.

One would not expect to find ice on Mercury, but close to the North and South Poles the crater floors are permanently in shade with temperatures as low as 125 K, so any ice existing there could remain for billions of years. It is thought that the ice, also later observed at the South Pole, is the remnant of comets that have impacted with Mercury in the past.

In 2004, a second NASA spacecraft left the Earth for Mercury. It made its first fly-by in January 2008 and will enter an elliptical orbit around it in 2011 after two further fly-bys. One might, perhaps, be surprised that it will have taken 7 years to enter an orbit around Mercury but it is harder to achieve this than one might

**Figure 3.5** Mercury as photographed by the Messenger spacecraft as it flew by on January 14, 2008. The colour has been slightly accentuated; younger surface material has a bluish tint. Image: MESSENGER, NASA, JHU APL, CIW.

think. Any spacecraft travelling to Mercury is, in some sense, 'falling' towards the Sun and gains considerable kinetic energy as it does so. Hence, the problem is not getting there, but slowing down sufficiently to be able to orbit it. It actually takes more energy to orbit Mercury than to escape from the Solar System. Europe and Japan are planning a joint mission to Mercury to be launched in 2013. The BepiColumbo spacecraft will reach Mercury in 2019.

## 3.5.2    *Venus*

| Mass | $4.8685 \times 10^{24}$ kg | 0.815 that of Earth |
|---|---|---|
| Radius | 6051.8 km | 0.95 that of Earth |
| Escape velocity | 10.46 km s$^{-1}$ | |
| Rotation period | −243 days (The minus sign indicates retrograde rotation.) | |
| Axial tilt | 177.36° | |
| Semi-major axis of orbit | 108 208 930 km | 0.7233 AU |
| Average orbital speed | 35.02 km s$^{-1}$ | |
| Eccentricity | 0.0068 | |
| Period | 224.7 days | |
| Orbital inclination | 3.39° | |

As Venus is seen either shining brightly in the East before dawn or, at other times, shining in the West after sunset, it once had two names. The 'evening star' was called Vesperus or Hesperus derived from the Latin and Greek words for evening, respectively, whilst the 'morning star' was called Phosphorus (the bearer of light) or Eosphorus (the bearer of dawn). It is said that the Greeks first thought that they were two different bodies but later came round to the Babylonian view that they were one and the same. There is a famous sentence in the philosophy of language 'Hesperus is Phosphorus' that implies an understanding of this fact.

Venus, which shines at close to magnitude $-4$, is the brightest object in the night sky after the Moon (Figure 3.6). As was shown in Chapter 1, Venus's angular size varies by a factor of about 5 as it orbits the Sun. However, when it is further away from the Earth and hence has the smaller angular size, a greater percentage of its surface is seen illuminated. The two effects tend to cancel out, with the brightness staying very close to magnitude $-4$ for several months at a time. Venus appears bright as it has a very high albedo, reflecting ~70% of the sunlight falling on it, the result of a totally cloud covered surface.

Twice, 8 years apart, every 120 years, Venus is seen to 'transit' across the face of the Sun. (In the twenty-first century in 2004 and 2012; see Figure 1.16.) Such transits were important historically as they gave a method of calculating the distance of Venus and hence, using Kepler's Third Law, of measuring the

**Figure 3.6** The surface of Venus as imaged by radar from the Magellan spacecraft. Image: Magellan Project, JPL, NASA.

astronomical unit. Captain Cook's exploration of Australia followed an expedition to Tahiti to observe the 1768 transit of Venus.

Much of our knowledge of Venus has come from observations by spacecraft. In December 1962 the Mariner 2 spacecraft flew over the surface at a distance of ~35 000 km and microwave and infrared observations showed that, whilst the cloud tops were very cool, the surface was at a temperature of at least 425°C. It found no evidence of a magnetic field.

The Russians made many, initially unsuccessful, attempts to land spacecraft on the surface. No one had anticipated an atmospheric pressure of ~100 times that of the Earth's. As a result, the descent parachutes were initially too large. This slowed the spacecraft's descent with the result that its batteries discharged before the craft reached the surface. Other craft were crushed by the great pressure in the lower atmosphere. Finally, in 1970, Venera 7 reached the surface sending temperature telemetry back for 23 min and then Vereras 9 and 10 sent back the first images of the surface revealing scattered boulders and basalt-like rock slabs (Figure 3.7). On their way to observe Halley's Comet in 1985 two Russian Vega craft sent balloon craft into the atmosphere which flew at an altitude of ~53 km above the surface and showed that there were high winds within a highly turbulent atmosphere.

Radar observations, initially from Earth, and then from an orbiting NASA spacecraft, Magellan, have given us detailed information about the surface structure. In 4.5 years, Magellan mapped 98% of Venus's surface with very high resolution. About 80% of the surface comprises smooth volcanic plains with the remainder forming highland regions, one in the northern hemisphere and one just south of the equator. The northern 'continent' is called Ishtar Terra, and is about the size of Australia and named after the Babylonian goddess of love. It rises to a peak of 11 km above the plains in the Maxwell Montes. The southern 'continent', called Aphrodite Terra after the Greek goddess of love, is somewhat larger having a surface area comparable with South America. The surface has impact craters, mountains and valleys along with a unique type of pancake-like, volcanic feature, called farras, which is up to 50 km in diameter and 1 km high.

**Figure 3.7** Basalt plain image from Venus by the Venera 14 lander. Image: Soviet Academy of Sciences.

The surface is thought to be ~500 million years old and, as a result, we see far more volcanoes than on Earth, where those older than ~100 million years will have been eroded. There are 167 volcanoes visible on Venus that are over 100 km across. The amount of sulphur dioxide in the atmosphere appears to be variable, which could indicate ongoing volcanic activity.

The atmosphere has been shown to largely consist of carbon dioxide with a small amount of nitrogen. Its mass is 93 times that of the Earth and results in an atmospheric pressure at the surface that is about 92 times that on the Earth. It contains thick clouds of sulphur dioxide and sulphuric acid may even 'rain' in the upper atmosphere! However, due to the surface temperature of over 460°C, this 'rain' would never reach the surface. This very high surface temperature is the result of the greenhouse gas effect of the very thick carbon dioxide atmosphere.

It is thought that Venus has a very similar internal structure to that of the Earth with a core, mantle and crust. The lack of plate tectonics – the relative movements of parts of the crust – may have prevented its interior cooling to the same extent as that of our Earth and it is thought that its interior is partially liquid. Venus is almost a twin to our Earth with a diameter just 650 km less and a mass 81.5% that of our planet but, once thought to have had an atmosphere much more like ours, evolution has taken it down a very different path!

### 3.5.3    *The Earth*

| | | |
|---|---|---|
| Mass | $5.9736 \times 10^{24}$ kg | |
| Radius (mean) | 6371 km | |
| Escape velocity | 11.18 km s$^{-1}$ | |
| Day length | 1 day | |
| Rotation period | 23 h 56 min 04 s (sidereal day) | |
| Axial tilt | 23.44° | |
| Semi-major axis of orbit | 149 597 887.5 km | 1 AU |
| Average Orbital Speed | 29.78 km s$^{-1}$ | |
| Eccentricity | 0.0167 | |
| Period | 365.256366 days | |
| Orbital inclination | 0° (Our orbit provides the reference plane.) | |

The Earth–Moon system was formed when an object, thought to have had a mass of about 10% that of the Earth, impacted with it soon after the formation of the Solar System (Figure 3.8). A portion of the combined mass was thrown off into space and formed the Moon, initially far closer to the Earth than now. As the

**Figure 3.8** Earthrise as seen by the Apollo 8 crew as they rounded the Moon's far side in December 1968. Image: NASA Apollo 8 Crew.

Moon's relative mass and size is comparable with that of the Earth the two are sometimes thought of as a 'double planet'.

The name Earth derives from the Anglo-Saxon word 'erda' meaning ground or soil. This became 'eorthe' in Old English and then 'erthe' in Middle English. Finally, from around 1400, the name Earth was used – the only planetary name not derived from Greek and Roman mythology.

Out-gassing from volcanoes produced its secondary atmosphere and water vapour condensed to form the oceans, with additional water provided by the impact of comets. About 4 billion years ago a self-replicating molecular system – life – arose. Photosynthesis allowed the trapping of solar energy and the by-product, oxygen, accumulated in the atmosphere. A resulting layer of ozone reduced the flux of ultraviolet radiation on the surface which allowed life forms to survive on land.

The surface of the Earth has been shaped by plate tectonics – the movement of sections of the crust across the underlying magma – which at times formed vast continents such as Pangaea which allowed species to colonize much of the surface.

In the Cambrian era, which followed a period of extreme cold when much of the Earth was covered in ice, multicellular life forms began to flourish. In the ~535 million years since then, there have been five 'mass extinctions' when many species died out. The last of these was 65 million years ago when a large asteroid or comet of at least 10 km in diameter formed the Chicxulub crater on (and offshore of) the Yucatan Peninsular in Mexico. The dust produced would have reduced the amount of sunlight reaching the ground and hence the growth of vegetation. This may have been one of the causes of the demise of the dinosaurs at about this time along with ~70% of all species then living on the Earth. Shrew-like small mammals were spared and their evolution finally gave rise to human beings.

Our Earth is a rocky body with a core, mantle and crust. Its rotation causes a bulge at the equator whose diameter is 43 km greater than the polar diameter. As a result, the peak of Mount Chimborazo in Equador lies furthest from the Earth's centre. With an extreme variation from 8.8 km above (Mount Everest) to 10.9 km below (the Mariana Trench), the surface of the Earth is actually smoother than a billiard ball!

The central core has a temperature of ~7000 K as a result of the decay of radio-active isotopes of potassium, uranium and thorium, all of which have half-lives of over 1 billion years. Convection currents within the molten rock brought this heat up towards the crust giving rise to thermal hotspots and produced the volcanic activity which gave the Earth its secondary atmosphere. The atmosphere is now largely composed of nitrogen (78%) and oxygen (21%) with the remaining 1% made up of water vapour, carbon dioxide, ozone, methane and other trace gases. It is worth pointing out again that without the warming due to the greenhouse gases, carbon dioxide, water vapour and methane, the average surface tempera-ture would be about $-18°C$ and life would almost certainly not exist.

Ancient peoples put the Earth at the centre of the universe and regarded the human race as special. As it was found that our Sun is one of billions of stars in our Galaxy, a principle of mediocrity arose. Planets like Earth were thought to have been very common and so advanced life like ours was thought to be widespread – we were not special. As we have learnt more about the history of our Earth, how plate tectonics have helped recycle carbon dioxide back into the atmosphere, how Jupiter prevents too many comets from impacting the Earth and how our large moon has stabilized the Earth's rotation axis, some scientists now believe that the conditions that have allowed intelligent life to arise here may well be very uncommon. It really could be that we *are* special, and it is not impossible that we are the only advanced life form now within our Milky Way Galaxy.

### 3.5.4    *The moon*

| Mass | $7.3477 \times 10^{23}$ kg |
|---|---|
| Radius (mean) | 1737.1 km |
| Rotation period | 27.3216 days (in synchronous rotation) |
| Semi-major axis of orbit | 384 399 km |
| Eccentricity | 0.0549 |
| Orbital period | 27.3216 days |
| Average orbital speed | 1.022 km s$^{-1}$ |
| Synodic period | 29.5306 days (from new moon to new moon) |
| Orbital inclination | 5.145° (with respect to the ecliptic) |

The Moon is the fifth largest satellite in the Solar System. It has a diameter slightly more than a quarter that of the Earth and its average distance is about 30 times that of the Earth's diameter. The gravitational pull on its surface is about one-sixth that on the Earth. Due to the fact that it has an elliptical orbit, its angular size varies by about ~12%; from 0.5548° at perigee, when it is closest to the Earth, down to 0.4923° at apogee, when it is furthest from the Earth. Partly due to its eccentric orbit, partly due to the inclination of its orbit and partly due to the fact that at moonrise and moonset we see it from different relative positions in space we can observe a total of 59% of the Moon's surface at one time or another. This effect is called libration.

The well known Moon illusion makes the Moon appear largest when near the horizon. However, it will of course be closer to us when highest in the sky and its angular size will actually be about 1.5% larger! Our perception of size is linked with how far away we believe an object is from us. It appears that we 'see' the celestial sphere above us, not as a true hemisphere, but one which is flattened overhead so that we believe that the objects above us in the sky are nearer to us than those near the horizon. So, observing the Moon above us, we believe it to be closer and mentally reduce its perceived size.

The Moon only reflects about 8% of the light incident upon it and is one of the least reflective objects in the Solar System reflecting about the same proportion of light as a lump of coal. The side of the Moon that faces Earth is called the near side, and the opposite side the far side. Even with the unaided eye one can clearly see that there are two distinct types of surface on the near side. We see light regions called 'highlands' and darker areas of the surface that we call 'maria', so-called because they were thought to be seas and oceans and were given beautiful names such as Oceanus Procellarum, Mare Tanquillatatis and Sinus Iridum – the Ocean of Storms, the Sea of Tranquillity and the Bay of Rainbows. When the far side of the Moon was first photographed by the Soviet probe Luna 3 in 1959, a surprising feature was its almost complete lack of maria.

We now know that the maria regions are covered by basaltic lava. Many are circular, such as Mare Crisium, which implies a number of giant impact events. These would have shattered the lunar crust allowing lava to well up and fill the resulting depressions. The fact that the maria are relatively lightly cratered compared with the highland regions shows that they were formed towards the end of the period in the early history of the Solar System when the surfaces of the planets and moons were being bombarded with the debris left over from its formation.

On the near side maria cover about 32% of the surface, but the far side has just a few patches making up just 2% in area. This requires an explanation and it appears that several factors may have contributed. Elements to produce heat by radioactive decay seem to be concentrated in the near-side hemisphere so that volcanic activity – producing the lava that filled the impact basins – would have been more prolific. The crust may also be thinner making it easier for the lava to have broken through to the surface. Perhaps less likely, is the fact that the Earth may have gravitationally 'captured' objects that might have impacted the near side and so destroyed the smooth maria regions.

The lighter-coloured regions of the Moon are commonly called highlands, since they are higher than most maria, though their formal name is Terrae. We should expect them to be higher. The basalt rocks that make up the maria are denser than the rocks making up the highland regions. For the Moon to be in hydrostatic equilibrium when at earlier times its interior was partially molten, at some constant depth below the surface the pressure must be equal. This implies that the column mass above this depth (the mass of, say, a 1 m diameter column to the surface) must be equal. This means that the columns of lighter material must be taller so the highland regions will rise above the maria.

Around the rims of the giant impact craters are seen several prominent mountain ranges – the surviving remnants of the impact basin's outer rims. Interestingly, four regions on the rim of the crater Peary, at the Moon's North Pole, are illuminated throughout the lunar day making this a possible site for a lunar base as solar panels could be used to provide a constant energy supply. This is due to the fact that the Moon has a very small tilt to the plane of the ecliptic. A further consequence of this fact is that regions at the bottom of craters near the Pole are in permanent shadow. This would have allowed ice, lodged there from cometary impacts, to have remained and thus be able to provide a source of water, oxygen and hydrogen. It should be noted that oxygen and hydrogen make an excellent rocket fuel so that perhaps a base on the Moon could be used to build and fuel rockets for further space exploration. It requires far less energy to launch a rocket from the Moon than from the Earth.

From radar observations made by the Clementine spacecraft it appears that in the craters near the Poles there may be pockets of ice making up a total of $1\,km^3$ but recent radar observations from Earth suggest that the radar signature indicative of water ice might, instead, be due to ejecta from young impact craters. The presence of significant water ice on the Moon has still to be proven.

Very obvious features of the Moon's surface are the lunar impact craters formed when asteroids and comets collided with the lunar surface. There are about half a million craters with diameters greater than 1 km. The largest is some 2240 km in diameter and 13 km in depth. Two prominent young craters seen on the near side are Tycho and Copernicus. Near full moon, light coloured rays of ejecta can be seen radiating from Tycho, so it is termed a rayed crater. These are visible in the lunar eclipse image shown in Figure 3.9.

The surface of the Moon is covered by what is called the lunar regolith. It has a thickness of about 3–5 m in the maria regions and 10–20 m in the

**Figure 3.9** The Moon seen at the time of total eclipse. Image: Ian Morison.

highland regions. Beneath the regolith lies a region of highly fractured bed-rock about 10–40 km deep. It is believed that, like the Earth, the Moon has a crust (about 50 km thick), mantle and core. When, following its formation the Moon was molten, the heavier elements fell towards its centre – a process called differentiation – to give a dense core. The core, composed largely of iron, is thought to be small with a radius of less than 350 km and is, at least partially, molten.

### Finding the height of a lunar mountain

The image of the third quarter Moon shown in Figure 3.10 was taken on a night when the 'seeing' was excellent and the resolution was of order 1 arcsec. Near the top of the image and close to the terminator (the line between the bright and dark sides of the Moon) is Mons Piton. This area is shown in greater detail on the right. Piton's shadow can be seen on the surface of the Moon. The length of the shadow is determined by the height of the mountain and the elevation of the Sun and so one can use this image to calculate the height of Mons Piton.

To simplify the calculation we will assume that the Sun is at infinite distance and above the Moon's equator (which runs across the face of the Moon at right angles to the terminator). Mons Piton lies at a lunar latitude of $+41°$. First, we need to calculate the radius of the circle of constant latitude round the surface of the Moon at $+41°$. This is a function of the lunar radius of 1738 km and the latitude of Piton.

*(continued)*

**Figure 3.10** The third quarter Moon and the region around Mons Piton.
Image: Ian Morison.

The radius of the circle of latitude at 41° north is given by $R_{moon} \times \cos(41°)$ and thus equals 1311.7 km.

Imagine slicing off the top of the Moon at the latitude of Piton and looking down on it from above as shown in Figure 3.11.

The lunar 'prime meridian' runs between Piton and the crater Cassini whose centre is on the terminator. Piton lies 1° west and Cassini 4.6° east of the prime

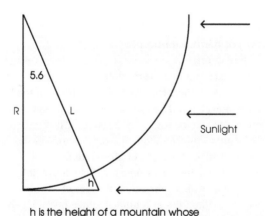

h is the height of a mountain whose
shadow would reach the terminator

**Figure 3.11** The geometry of the Mons Piton height calculation.

meridian. The angle subtended by the separation of Piton and Cassini at the centre of this circle is thus $4.6° + 1° = 5.6°$.

From the radius calculated above it is possible, using the geometry of Figure 3.11, to calculate the height that Piton would have in order to produce a shadow that just reached the terminator. The height $h$ of a mountain located at the site of Piton whose shadow would just reach the terminator is given by $L - R$

$$R/L = \cos 5.6°$$

So

$$L = R/\cos 5.6 = 1311.7/0.9952 = 1318.0$$

Thus

$$h = L - R = 1317.1 - 1311.7 = 6.3\,\text{km}$$

Piton obviously has a lesser height than this and, from the lunar image, it is possible to estimate its height from the ratio of the length of its shadow to the distance from the mountain to the terminator. From an image such as that to the right of Figure 3.11 we can measure the length of the Piton's shadow ($l_1$) and the distance of Piton from the terminator ($l_2$). Using a large scale image, $l_1$ was measured to be ~12 mm and $l_2$~33 mm.

Thus an estimate of the height of Mons Piton is given by:

$$6.3 \times 12/33 = 2.29\,\text{km}$$

This estimate is close to the actual value of 2.25 km.

## Tides

Tides in the oceans are the result of the inverse square law of gravity and the size of the Earth. The gravitational force on the nearest side of the Earth to the Moon is thus greater than at the mid point and even more so than that at the far side. This causes a differential gravitational effect called a tidal force whose effects fall off as the fourth power of the distance. The result is to stretch out the Earth's oceans into an ellipse so that the sea level is higher closest to the Moon (as might well be expected) but also higher on the far side (less obvious) as the force on the oceans there is less than at the midpoint. As the Earth spins on its axis, these two bulges rotate around the Earth so we normally get two high tides per day. As the Moon is also moving around the Earth in a period not too far off 24 days the two high tides will appear earlier by about 1 h per day.

The Sun causes a tidal force as well, but its effects are less (at about 46%) than that of the Moon. At new and full moon, the tidal forces reinforce to give what are called spring tides whilst at first and third quarter they partially cancel giving neap tides which have a reduced tidal range. As both the Moon and Earth are in elliptical orbits the height of the spring tides can vary significantly. When the Moon is at perigee (nearest the Earth) the tides will be higher whilst, if this is the case when the Earth is at perihelion (nearest the Sun) in winter, the tidal forces are greatest and we get the highest tides.

The gravitational coupling between the Moon and the oceans affects the orbit of the Moon. Due to the Earth's rotation, the tidal bulges do not point directly towards (and away from) the Moon. Gravitationally, the mass of the Earth could be regarded as made up of three parts:

(1)    A mass at the centre of the Earth representing that part (the vast majority) which is spherically symmetric.
(2)    One representing the tidal bulge on the near side of the Earth to the Moon which is *ahead* of the Earth–Moon line.
(3)    One on the far side of the Earth which is *behind* the Earth–Moon.

The gravitational pull on the Moon of the nearer water mass is greater. As this is leading the Earth–Moon line its effect is to try to advance the Moon in its orbit. The overall effect is to transfer angular momentum from the Earth's rotation to the Moon. As a result, the Moon is moving to a higher orbit with a longer period and, each year, the separation of the two bodies is increasing by about 3.8 cm.

### Lunar eclipses

If the plane of the Moon's orbit was the same as that of the Earth's orbit around the Sun, we would get eclipses of the Sun at new moon and eclipses of the Moon at full moon. Due to the fact that the Moon's orbit is inclined by about 5°, the Moon often passes above or below the Sun–Earth line, and eclipses happen less often. It might be thought that the Moon would totally disappear during a lunar eclipse, but light that is scattered through the Earth's atmosphere falls on the Moon so it can still be seen – though far less bright. If we observed the Earth from the Moon during a lunar eclipse we would see that it would have a red limb as the molecules in the Earth's atmosphere scatters blue light more than red (which also explains why the Sun appears red at sunset). The Moon thus takes on a reddish hue at totality with the brightness and colour very dependant on the amount of dust in the Earth's atmosphere. Following a major volcanic eruption, so little light reaches the Moon it can appear a very dull dark grey but conversely, when the atmosphere is relatively free of dust, it can appear a beautiful orange red (Figure 3.9).

### Lunar exploration

The Moon has been studied more than any other body in the Solar System: it has been imaged from above by lunar orbiters, and its surface studied by a number of landers – the first being Lunar 9 in 1965, followed by the Russian lunar rovers and the NASA Surveyor craft. Lunar exploration culminated in the NASA Apollo programme when six spacecraft landed men on the Moon. Samples of Moon rocks have been brought back to Earth by three Russian missions (Luna 16, 20, and 24) and the Apollo missions 11, 12, 14, 15, 16 and 17.

As shown in Figure 3.12, the Apollo missions left science packages to measure heat flow, magnetic fields and seismic oscillations along with corner-cube light reflectors. Each reflector assembly contained one hundred reflecting elements similar to 'cat's eyes'. The light from laser equipped telescopes on Earth is reflected to enable the distance from the telescope to the Moon to be measured to an accuracy of less than a centimetre. These have enabled the Moon's orbit to be measured with very high precision and, one might point out, provides irrefutable evidence that the Apollo missions *did* go to the Moon! There is an interesting aspect of diffraction theory related to these reflectors which will be analysed in Chapter 5. If the reflectors were perfect, the light would be reflected directly back to the point where the laser pulse had been transmitted. The problem is that the pulse would return

**Figure 3.12**  Tranquillity Base. The Eagle Lander lies behind the seismograph. The data that it acquired were transmitted back to Earth using a cylindrical antenna. Behind this antenna, at right angles to the Earth, is the Luna Laser Reflector. Image: Neil Armstrong, Apollo 11 Crew, GRIN, NASA.

there 2.5 s later by which time the telescope on the Earth that had transmitted the pulse would have travelled some way around the globe. It would no longer be in a position where it could receive the returned laser pulse! The use of multiple small reflectors spreads the returned beam so that the pulse can be detected.

Following a 25 year lull in lunar exploration, in recent years spacecraft from Europe, the USA, and China have returned to orbit the Moon with a Russian orbiter and lander planned for 2012. NASA is now making plans for a manned return to the Moon by the year 2020 and the building of a permanent base near one of the poles.

### 3.5.5    *Mars*

| Mass | $6.4185 \times 10^{23}$ kg | 0.107 that of Earth |
|---|---|---|
| Radius (equatorial) | 3.396 km | 0.533 that of Earth |
| Radius (polar) | 3,376 km | 0.531 that of Earth |
| Escape velocity | 5.027 km s$^{-1}$ | |
| Rotation rate | 1.026 days | |
| Axial tilt | 25.19° | |
| Semi-major axis of orbit | 227 936 637 km | 1.52 AU |
| Average orbital speed | 24.077 km s$^{-1}$ | |
| Eccentricity | 0.0934 | |
| Orbital period | 686.96 days | 1.8808 Julian years |
| Orbital inclination | 1.85° | |

Mars is often called the 'red planet' but, to the author's eyes, appears more of a salmon pink (Figure 3.13). It is a rocky planet having about half the diameter of the Earth but only one-tenth its mass. The reddish tint is due to oxides of iron on the surface known as haematite or rust which forms a dust with the consistency of talcum powder. Its atmosphere is very thin, about one-hundredth that of Earth's, and is largely composed of carbon dioxide (95%) along with nitrogen (3%), argon (1.6%) and traces of water vapour and oxygen.

As the Earth and Mars have similar axial tilts, they have comparable seasons but Mars season lengths are about twice those of Earth as the Martian year is approximately two Earth years in length. Martian surface temperatures range from approximately −140°C during winter up to 20°C in summer. Mars also suffers from dust storms which can occasionally cover the planet's entire surface. It has two polar ice caps which are primarily composed of water ice but covered by a layer of solid carbon dioxide (dry ice). At the South Pole the carbon dioxide

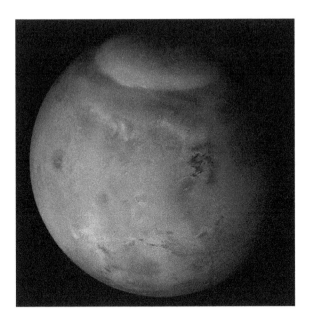

**Figure 3.13** Mars in spring time. At the bottom right is Valles Marinaris, a canyon four times as deep and three times as long as the Grand Canyon on Earth. On the left are several volcanoes including Olympus Mons, a volcano three times higher than Mt Everest. At the top is the north polar cap made of thawing water ice and 'dry ice' (solid carbon dioxide). Image: MSSS, JPL, NASA.

layer is ~8 m deep overlying ~3 km of water ice within a diameter of ~350 km. The north polar cap has a diameter of ~1000 km and is about 2 km thick. During the winter, dry ice builds up a ~1 m layer above the water ice and this causes a reduction of carbon dioxide in the atmosphere, so reducing the atmospheric pressure.

### Was there a civilization on Mars?

Mars was first seen through a telescope by Galileo in 1609, but his small telescope showed no surface details. When Mars was at its closet to Earth in 1877, an Italian astronomer, Giovanni Schiaparelli, used a 22 cm telescope to chart its surface and produce the first detailed maps. They contained linear features which Schiaparelli called canali, the Italian for channel. However this was translated into English as 'canal' – which implies a man-made water course – and the feeling arose that Mars might be inhabited by an intelligent race. It should be pointed that a waterway *could not* be detected from Earth, but it was thought that these would have been used for irrigation and so would have irrigated crops growing adjacent to them which *could* be seen from Earth.

Influenced by Schiaparelli's observations, Percival Lowell founded an observatory at Flagstaff, Arizona (later famous for the discovery of Pluto) where he made detailed observations of Mars which showed an intricate grid of canals. However, as telescopes became larger, fewer canali were seen though the surface showed distinct features. They appear to have been an optical illusion but the myth of advanced life on Mars was not finally dispelled until NASA's Mariner spacecraft reached Mars in the 1960s. (An intriguing image of a huge rock formation, called 'the face on Mars', was obtained by the Viking 1 spacecraft in 1961 leading some to suspect that this was a giant representation of a past civilization. However, a detailed photograph taken by Mars Global Surveyor in 2001 showed it to be a 'mesa' – a broad, flat topped rock outcrop with steep sides.)

The detailed images taken by the Mariner spacecraft showed giant canyons and vast volcanoes. One of these, Olympus Mons, is the largest known volcano in the Solar System with a caldera of 85 km in width surmounting the volcanic cone whose base is 550 km in diameter. The caldera is nearly 27 km above the Martian surface, three times higher than Everest! It was realised that when these giant volcanoes were active, some 3000 million years ago, they would have given Mars a far thicker atmosphere than now and the effects of greenhouse gases in the atmosphere would have allowed the surface temperature to be sufficiently high for water to exist on the surface. Other visible surface features gave ample evidence of water flow over the surface leading to speculation that simple life forms might then have existed on Mars.

This possibility led to two Viking Landers being sent to Mars in 1976. As well as imaging the surface and collecting scientific data, their objective was to search for any evidence of life. They conducted three experiments which, though discovering unexpected chemical activity in the Martian soil provided no clear evidence for the presence of any living organisms. As Mars has a very thin atmosphere (and no ozone layer), far more ultraviolet light reaches the surface than on Earth. This would prevent the existence of life above ground. If life had once arisen on Mars, one could now only expect to find evidence for it beneath the surface.

Since then, many spacecraft have reached Mars. Two rovers, Spirit and Opportunity, have roamed across Mars and provided proof that water once existed on the surface (Figure 3.14). Only one lander, Beagle 2, has attempted to directly search for evidence of simple life since the 1970s but, sadly, it appeared to have crash landed on Christmas Day 2003.

Mars has two moons, Phobos and Deimos, which may be captured asteroids (see next section). Phobos is a small, irregularly shaped object about 22 km in diameter (Figure 3.15). It orbits Mars at a height of 9377 km, closer than any other satellite to its planet and moves so fast around Mars that it actually rises in the west and sets in the east. Deimos is just 6 km across in a nearly circular orbit.

**Figure 3.14** A Martian panorama imaged by the the Spirit Rover at Gusev Crater.
Image: Mars Exploration Rover Mission, Cornell, JPL, NASA.

**Figure 3.15** Phobos imaged by the Mars Reconnaissance Orbiter in March 2008.
Image: HiRISE, MRO, LPL (U. Arizona), NASA.

It takes just over 30 h to complete one orbit, not that much greater than Mars's rotation period so it will remain visible for ~2.7 days between rising and setting. It would appear star like as it subtends an angle of just 2.5 arcsec and is about as bright as the planet Venus appears to us.

### 3.5.6    *Ceres and the minor planets*

| Mass | $9.46 \times 10^{20}$ kg | |
|---|---|---|
| Radius (equatorial) | 487 km | |
| Radius (polar) | 455 km | |
| Rotation period | 0.378 days | |
| Semi-major axis of orbit | 414 703 838 km | 2.77 AU |
| Orbital period | 1679.8 days | |
| Eccentricity | 0.0797 | |
| Orbital inclination | 10.59° | |

In 1768, Johann Elert Bode suggested that there might be a planet in orbit between Mars and Jupiter. He based this on the so-called Titius–Bode Law (see box), an empirical law proposed by Johann Daniel Titius in 1766 which (roughly) gave the relative orbital distances of the planets from the Sun. The Law 'predicted' that a planet should lie at a distance from the Sun of 2.8 AU and, with the discovery some years earlier of Uranus which fitted the Law's predictions quite well, a group of 24 astronomers, who became known as the 'Celestial Police', combined their efforts to make a methodical search for the possible planet. Though they did not discover Ceres that did find several minor planets – also known as asteroids—in what is now called 'the main asteroid belt'.

The body which became known as Ceres was discovered on New Year's Day 1801 by Guiseppe Piazzi who first thought it to be a comet. Ceres has a diameter of ~950 km, and has sufficient mass to be in hydrostatic equilibrium (making it round) and so has become the smallest of the dwarf planets. It contains almost one-third of the mass of the main belt asteroids of which over 170 000 now have computed orbits. In 2006 the IAU classed these as Small Solar System Bodies.

It is though that Ceres is sufficiently large to have become differentiated, that is, the heavier rocky elements have concentrated in the centre giving Ceres a core over which lies an ice mantle (a mixture of water ice and minerals such as carbonates and clay).

**The Titius–Bode law**
If one takes the number 0.3, continuously doubles it and adds 0.4 to each value one gets a sequence of values which closely agree with the distance of the planets in AU. (Zero is used as an initial value.)

| 0.4 | 0.7 | 1.0 | 1.6 | 2.8 | 5.2 | 10 | 19.6 |

The planets then known fell at distances:

| 0.39 | 0.72 | 1.0 | 1.52 | ? | 5.2 | 9.5 | 19.2 |

This suggested that a planet might exist at a distance of ~2.8 AU.

### Near-Earth Objects

Through gravitational interaction within the main belt of asteroids, occasionally an asteroid can acquire an orbit which will bring it into the inner Solar System. Should they come within the orbit of the Earth, they have the potential to impact the Earth's surface. Such asteroids, along with comets and meteoroids (<50 m across), which come close to the Earth are termed Near-Earth Objects (NEOs). They are usually only spotted as they pass close to the Earth. NASA has a mandate to find all NEOs that have a diameter greater than 1 km, as these have the potential to cause catastrophic local damage and even global effects. By the end of 2007, approximately 800 above this size had been detected out of an expected 1000. Nearly 5000 NEOs have been detected including over 60 near-Earth comets. Of these, around 800 have been classified as potentially hazardous. The hope is that should a NEO be discovered that might impact the Earth at some time in the future it would be possible to alter its orbit sufficiently to miss the Earth.

In June 1908, a giant explosion, 1000 times greater than the Hiroshima atomic bomb, occurred close to the Tungusta river in northern Siberia. It is though that a comet or asteroid of about 50 m diameter exploded above the ground – felling over 800 million trees in an area of over 2000 km². If the impact had occurred 4 h and 47 min later, the city of St Petersburg (Leningrad) would have been destroyed. It is estimated that there are over 200 000 NEOs of comparable size and the probability of an impact of any one with Earth is of the order of one in 100 million per year. One such impact might thus be expected every 500 years so that there is a 1 in 500 chance of an impact in any one year. It is quite likely that such an impact would be into the oceans causing a tsunami which would have repercussions over a significant part of the globe.

As of 2008, the most potentially serious threat to Earth is from Asteroid (29075) 1950 DA. It has a diameter of ~1 km and a 1 in 300 chance of hitting the Earth on March 16, 2880.

## 3.5.7    *Jupiter*

| Mass | $1.9 \times 10^{27}$ kg | 317.8 times that of Earth |
|---|---|---|
| Radius (equatorial) | 71.492 km | 11.21 times that of Earth |
| Radius (polar) | 66 854 km | 10.52 times that of Earth |
| Rotation period | 9.925 h | |
| Axial tilt | 3.13° | |
| Semi-major axis of orbit | 778 547 199 km | 5.2 AU |
| Orbital period | 4334.5 days | |
| Average orbital speed | 13.07 km s$^{-1}$ | |
| Eccentricity | 0.049 | |
| Orbital inclination | 1.3046° | |

With Saturn, Uranus and Neptune, Jupiter is one of the gas giants of the Solar System and its mass exceeds that of all the other planets combined by two and a half times. Its interior mass is primarily made up of hydrogen (~71%) and helium (24%) with ~5% of heavier elements. Its composition thus closely follows that of the solar nebula from which it was formed. Interestingly, if Jupiter were to acquire more mass, its diameter would actually decrease, so it is about as large as a planet of its composition could be.

Jupiter is thought to consist of a dense core surrounded by a layer of liquid metallic hydrogen lying under an outer layer, about 1000 km thick, composed very largely of molecular hydrogen. Jupiter is perpetually covered with a cloud layer about 50 km thick. The clouds are composed of ammonia crystals arranged into bands of different latitudes made up of light coloured zones between darker belts. The orange and brown colours in the Jovian clouds are caused by compounds containing phosphorus and sulphur exposed to ultraviolet light from the Sun. At differing latitudes, the darker clouds so formed deeper within the atmosphere are masked out by higher clouds of crystallizing ammonia producing the pale zones seen between the belts (Figure 3.16).

### The Great Red Spot

Wind speeds of up to 100 m s$^{-1}$ are common in the atmosphere and opposing circulation patterns caused, in part, by Jupiter's rapid rotation rate cause storms and turbulence in the atmosphere. The belts and zones are seen to vary in colour and form from year to year, but the general pattern remains stable. The best known feature in the atmosphere is undoubtedly the Great Red Spot. It is a persistent anticyclonic storm, more than twice the diameter of the Earth, which has been observed

**Figure 3.16** Jupiter as imaged by the Hubble Space Telescope showing the Equatorial Bands and the Great Red Spot below and to the left of which is a new feature nicknamed 'Red Spot Junior'. Image: NASA, ESA, A. Simon-Miller (Goddard Space Flight Centre), I. de Pater, M. Wong (UC Berkeley).

since at least 1831. It rotates in an anticlockwise direction with a rotation period of about 6 days and is thought to be stable and so has become a permanent, or at least a very long term feature of the Jovian atmosphere. It is not, however, fixed in position, and though staying at latitude 22° south has moved around the planet several times since it was first observed. Similar, but smaller, features are common, with white ovals of cool clouds in the upper atmosphere and warmer brown ovals lower down. These smaller storms can sometimes merge to form larger features, as happened in 2000 when three white ovals, first observed in 1938, combined into one. In the following years its colour has reddened and it has been nicknamed 'Red Spot Junior'.

### The rings of Jupiter

Jupiter has a very faint planetary ring system composed of three main segments: an inner halo, a brighter main ring, and an outer 'gossamer' ring having two distinct components. They appear to be made of dust with the main ring probably made of material ejected from the satellites Adrastea and Metis as a result of meteorite impact. Jupiter's strong gravitational pull prevents the material falling back onto their surfaces and they gradually move towards Jupiter. It is thought that the two components of the gossamer ring are produced in similar fashion by the moons Thebe and Amalthea.

### The Comet Shoemaker-Levy 9 impact

Perhaps the most exciting event in recent Jovian history was when the fragments of Comet Shoemaker-Levy 9 impacted on its surface in July 1994 (Figure 3.17). It is thought that Jupiter had captured the comet in the late 1960s or early 1970s so it had become a temporary satellite of Jupiter orbiting it once every 2 years or so. It appears that the comet had passed within 40 000 km of Jupiter's surface on July 7, 1992. This distance is within what is called the Roche limit of the planet within which the gravitational forces acting on the comet were able to break it up into a number of fragments. If a body has a particular size, the force of gravity acting on those parts of the body furthest from the centre of mass of a nearby planet will be less than that on the nearer parts. A differential 'tidal' force thus acts across the body. If this force is greater than the gravitational forces keeping the body whole, the body will break up into smaller parts. As Jupiter is very massive, its tidal forces are very great and also have a significant effect on Jupiter's innermost Galilean moon, Io.

The comet was discovered on the night of March 24, 1993 by Carolyn and Eugene Shoemaker and David Levy in a photograph taken with the 0.4 m Schmidt

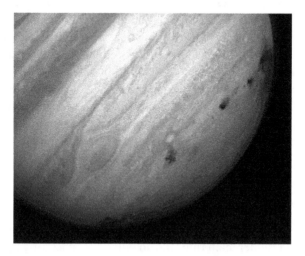

**Figure 3.17** The impact sites where fragments of the Comet Shoemaker-Levy 9 impacted on the surface of Jupiter. Image: Hubble Space Telescope, Space Telescope Science Institute, NASA.

Telescope at the Mount Palomar Observatory in California. (They were conducting a programme of observations designed to uncover NEOs.)

High resolution images, such as that made by the Hubble Space Telescope, showed a number of fragments (labelled A to W) which ranged in size from a few hundred metres up to 2 km. It is thought that the initial diameter of the comet would have been about 5 km. As the orbits of the fragments were refined, it became apparent that they were very likely to impact on the surface of Jupiter the following year with the individual impacts to be spread out over just under a week.

Further observations showed that the impact sites would lie just over Jupiter's limb, so would not be directly visible from Earth. However, in the same way that a nuclear explosion creates a giant mushroom cloud that rises up into the atmosphere, the Hubble Space Telescope was trained on the limb in the hope of observing a fireball above the impact site. The spacecraft Galileo, en route to Jupiter, was in a position to image the impact site and on July 16, 1994 detected a fireball which had a peak temperature of 24 000 K. Observations from Earth soon detected the plume from the fireball that reached over 3000 km above the Jovian atmosphere.

The material thrown up into space soon fell back to the surface and formed dark spots, similar in size to the Earth, on the surface. They were easily visible with a small telescope and the author will never forget his first sight of the pair of giant spots produced by the F and G fragments. The latter produced a spot some 12 000 km across and is thought to have released the equivalent of 6 000 000 t of TNT! Over 6 days, 21 discrete impacts were observed (the last on July 22 when fragment W struck the planet). The visible effects of these impacts highlighted the effects such events would have should the Earth, not Jupiter, be the target and spurred on the programmes that were already in place to detect those asteroids that might be a threat to the Earth.

## Jupiter's Galilean moons

Even a very small telescope can detect the four major moons of Jupiter as they weave their way around it. In order of distance from Jupiter, they are called Io, Europa, Ganymeade and Callisto and are comparable in size with our Moon (Figure 3.18). Discovered by Galileo in 1610, they showed him that Solar System objects did not all have to orbit the Sun, giving further evidence for the Copernican model of the Solar System.

Observations in 1676 made by the Danish astronomer Christensen Roemar of the times of their eclipses as they passed behind Jupiter led to the first determination of the speed of light. An eclipse of Io occurs every 42.5 h – the period of its orbit – and it thus provides a form of cosmic clock. However, Roemar

**Figure 3.18** Jupiter's moon Io showing volcanoes and sulphur dioxide frost.
Image: Galileo Project, JPL, NASA.

observed that the 40 orbits of Io during the time that the Earth was moving towards Jupiter took a total of 22 min less than when the Earth was moving away from Jupiter ~6 months later. The change in apparent period is due to the Doppler effect and this enabled him to calculate the ratio of the velocity of light to the orbital speed of the Earth around the Sun. He derived a value for this ratio of ~9300. As the orbital speed of the Earth is ~30 km s$^{-1}$ this gave a value (actually calculated by Christiaan Huygens from Roemar's observations) for the speed of light of ~279 000 km s$^{-1}$.

The two innermost moons, Io and Europa, are of great interest. Io is the fourth largest moon in the Solar System with a diameter of 3642 km. When high resolution images of Io were received on Earth from the Voyager spacecraft in 1979, astronomers were amazed to find that Io was pockmarked with over 400 volcanoes. It was soon realised that giant tidal forces due to the close proximity of Jupiter would pummel the interior, generating heat and so give Io a molten interior. As a result, in contrast with most of the other moons in the outer Solar System which have an icy surface, Io has a rocky silicate crust overlying a molten iron or iron sulphide core. A large part of Io's surface is formed of planes covered by red and orange sulphur compounds and brilliant white sulphur dioxide frost. Above the planes, are seen over 100 mountains, some higher than Mt Everest – a strange world indeed.

In contrast, Europa, the sixth largest moon in the Solar System with a diameter of just over 3000 km, has an icy crust above an interior of silicate rock overlying a probable iron core. The icy surface is one of the smoothest in the Solar System. Close up images show breaks in the ice as though parts of the surface are breaking apart and then being filled with fresh ice. This implies that the crust is floating above a liquid ocean, warmed by the tidal heating from its proximity with Jupiter. This could thus conceivably be an abode for life and some ambitious proposals have been made for a spacecraft to land and burrow beneath the ice to investigate whether any life forms are present!

### 3.5.8    *Saturn*

| Mass | $5.68 \times 10^{26}$ kg | 95.15 times that of Earth |
|---|---|---|
| Radius (equatorial) | 60 268 km | 9.45 times that of Earth |
| Radius (polar) | 54 364 km | 8.55 times that of Earth |
| Rotation period | 10 h 32 min to 10 h 47 min | |
| Axial tilt | 26.73° | |
| Semi-major axis of orbit | 1 433 449 370 km | 9.58 AU |
| Orbital period | 10 832.327 days | 29.66 years |
| Average orbital speed | 9.69 km s$^{-1}$ | |
| Eccentricity | 0.056 | |
| Orbital inclination | 2.485° | |

Galileo first observed Saturn with his telescope in 1610 and became somewhat perplexed. He described the planet as having 'ears' and composed of three bodies which almost touched each other with that at the centre about three times the size of the outer two whose orientation was fixed. He became even more perplexed when 2 years later the outer two bodies had gone. 'Has Saturn swallowed his children?' he wondered. He became further confused when they reappeared in 1613. In 1655 Christiaan Huygens observed Saturn with a far superior telescope and suggested that Saturn was surrounded by a ring system. He wrote: 'Saturn is surrounded by a thin, flat, ring, nowhere touching, inclined to the ecliptic'.

As telescopes improved, more details could be seen and, in 1675, Giovanni Domenico Cassini observed that Saturn's ring system was composed of a number of smaller rings separated by gaps the largest of which has become known as 'Cassini's Division'. In the mid 1800s, James Clerk Maxwell showed that a solid ring could not be stable and would break apart so that the ring system must be made up of myriads of particles individually orbiting Saturn. This would imply that different annuli of the rings would be moving at different speeds around Saturn and this was proved when James Keeler of the Lick Observatory made spectroscopic studies of the ring system in 1895.

There is no doubt that, due to its ring system, Saturn is the most beautiful object in the Solar System that can be observed with a small telescope (Figure 3.19). The key to understanding Galileo's confusion lies in Huygens's description that the ring system was inclined to the ecliptic due to Saturn's axial tilt. Assume that Saturn's North Pole was, at some point in its orbit, tilted closest to the Sun. Close to the Sun we, on Earth, would see much of the northern hemisphere and the rings at their most open. Just under 15 years later, Saturn will be on the opposite side of its orbit and the North Pole would be tilted away from the Sun. We would then see the southern hemisphere best and the rings would also be wide open. Half way in between these extremes we see the rings edge-on and, just as Galileo observed, they effectively disappear. Hence, the Earth will lay in the ring plane twice every orbit, about once every 15 years.

It is not surprising that the rings effectively disappear as it is thought that they are less than 1 km in thickness! The ring particles range in size from dust particles up to boulders a few metres in size and are largely composed of water ice (~93%) along with amorphous carbon (~7%). Three rings can be observed from Earth that extend from 6630 to 120 700 km above Saturn's equator. The outer ring, A ring, has a significant gap within it, called Enkes Division, whilst Cassini's Division separates the A from the middle B, or Bright Ring. Two further rings have

**Figure 3.19** Saturn observed by the Cassini spacecraft as Saturn eclipsed the Sun. The far side of Saturn from the Sun is partially lit by light reflected from the rings. Image: CICLOPS, JPL, ESA, NASA.

been discovered more recently; within the C ring there is a very faint D ring, whilst outside the A ring is a very thin F ring.

The rings are thought to have been formed when a moon either came within the Roche limit of the planet where tidal forces broke it apart, or was impacted by a large comet or asteroid to give the same result. (As a small body nears a massive one, the gravitational force on the nearer side of the body exceeds that on the far side. There is thus a differential force across the body which tends to pull it apart. The Roche limit is the distance from a planet at which this force would break up a typical small body.)

The structure we see within the rings is due to the cumulative effect of the gravitational pull of Saturn's many moons. Where a moon has a period which is a simple multiple of that of particles at a certain (nearer) distance to the centre of Saturn, a 'resonance' occurs which clears particles from that part of the ring system. In this way, the moon Mimas clears particles from the Cassini Division.

### Titan

Titan is the largest moon of Saturn and the only the moon in the Solar System known to have a dense atmosphere. It is also the only object other than Earth for which there is evidence of surface liquids, in the form of hydrocarbon lakes, in the satellite's polar regions. It is about 50% larger than our Moon and 80% more massive, and is second only to Jupiter's moon Ganymede in size and larger (but not as massive) as Mercury. Like our Moon, it is tidally locked and always presents the same face to Saturn. Titan has a relatively smooth crust composed of water ice which overlays a rocky interior. The atmosphere is quite dense, largely made up of nitrogen (~98%) giving a surface pressure of more than one and a half times that of the Earth. Within the atmosphere are clouds of methane and ethane and an orange haze made up of organic molecules that result from the break up of methane in the atmosphere by ultraviolet light from the Sun. The source of this methane is somewhat of a mystery, as the Sun's ultraviolet light should eliminate methane from the atmosphere in about 50 million years. It is not impossible that it has a biological origin!

Observations, first from the Hubble Space Telescope and then from the Cassini spacecraft (in infrared light to observe through the haze) show that Titan's surface is marked by broad swaths of bright and dark terrain. The largest feature is Xanadu, about the size of Australia. A sophisticated radar system carried by the Cassini spacecraft showed evidence for hydrocarbon seas, lakes and a network of tributaries near the North Pole so finally confirming the presence of liquid methane on the surface.

### The Huygens probe

On December 25, 2004 the Huygens probe separated from the Cassini Orbiter that had carried it to Saturn and, having been lowered through its atmosphere by parachute, landed on the surface of Titan on January 14, 2005 (Figure 3.20). Images of the surface taken from a height of ~16 km showed what are considered to be drainage channels in light coloured higher ground leading down to the shoreline of a darker sea or plain. Some of the photos even seemed to suggest islands and a mist shrouded coastline. There was no evidence of any liquids at the time of landing, but strong evidence of its presence in the recent past.

As the spacecraft landed, a penetrometer studied its de-acceleration. It was initially thought that the surface had a hard crust overlaying a sticky material. One scientist compared the colour and texture of the surface with that of a crème brûlée, and another, with stepping on a cowpat! However, it may have been that the craft landed on, and then displaced, a pebble on the surface giving the effect of a surface crust and the surface may, in fact, consist of 'sand' made up of ice grains forming a flat plain covered with pebbles made of water ice.

**Figure 3.20** A panorama of the surface of Titan taken as the Huygen's probe was descending through its atmosphere. Image: ESA, NASA, Descent Imager/Spectral Radiometer Team (LPL).

### 3.5.9    *Uranus*

| Mass | $8.68 \times 10^{25}$ kg | 14.536 times that of Earth |
|---|---|---|
| Radius (equatorial) | 25 560 km | |
| Radius (polar) | 24 973 km | |
| Rotation period | 0.378 days | |
| Axial tilt | 97.77° | |
| Semi-major axis of orbit | 2 876 697 082 km | 19.23 AU |
| Orbital period | 30 799 095 days | 84.32 years |
| Average orbital speed | 6.81 km s$^{-1}$ | |
| Eccentricity | 0.0444 | |
| Orbital inclination | 0.772° | |

Uranus was the first planet to have been discovered in modern times and though, at magnitude ~5.5, it is just visible to the unaided eye without a telescope it would have been impossible to show that it was a star rather than a planet, save for its slow motion across the heavens. Even when telescopes had come into use, their relatively poor optics meant that it was charted as a star many times before it was recognised as a planet by William Herschel in 1781. John Flamsteed had observed it at least six times in 1690 and given it the star designation 34 Tauri whilst the French astronomer, Pierre Lemonnier, observed Uranus at least 12 times between 1750 and 1769.

William Herschel had come to England from Hanover in Germany where his father, Isaac, was an oboist in the band of the Hanoverian Foot Guards. As well as giving his third child, Freidrich Wilhelm Herschel, a thorough grounding in music he gave him an interest in the heavens. When 15 years old, William entered the band as an oboist and violinist and first came to England in 1755 when the Foot Guards were sent to help defend it from a feared French invasion. He soon returned to Hanover but in 1757, as the city was overrun by the French, he and his brother Jacob came to England for a second time. Jacob returned to Hanover to take up a position in the court orchestra but his brother, who had anglicized his Christian names to Frederick William, stayed to take up the position of instructor to the band of the Durham Militia.

Following a short period as a concert manager in Leeds he was offered the post of organist at the Octagon Chapel in Bath where he set up home. After some years he was able to persuade his family to allow his sister Caroline to come to Bath to act as his housekeeper. She had been acting as a servant in her father's house and William had to pay his father an allowance so that he could pay for a replacement!

She repaid William's kindness with great devotion, giving up a career as a singer in order to assist him, and later became a significant astronomer in her own right.

William's interest in astronomy increased and he became very unsatisfied with the telescopes that he could buy so decided that he would make a reflecting telescope of his own. In those days, the mirrors were cast in speculum metal – an alloy of two parts copper and one of tin – and he made the castings in the kitchen of his house. Occasionally the mould would break and the molten metal would crack the stone floor.

By 1778, he had built an excellent telescope having a mirror of just over 6 in. (150 mm) in diameter and began to make a survey of the whole sky. On the night of March 13, 1781 he observed an object that did not have the appearance of a star and he first thought that it was a comet:

'The power I had on when I first saw the comet was 227. From experience I know that the diameters of the fixed stars are not proportionally magnified with higher powers, as planets are; therefore I now put the powers at 460 and 932, and found that the diameter of the comet increased in proportion to the power, as it ought to be, on the supposition of its not being a fixed star, while the diameters of the stars to which I compared it were not increased in the same ratio. Moreover, the comet being magnified much beyond what its light would admit of, appeared hazy and ill-defined with these great powers, while the stars preserved that lustre and distinctness which from many thousand observations I knew they would retain.'

Observations over the following months showed that it did not have the highly elliptical or parabolic orbit of a typical comet and that it was in a nearly circular orbit having a semi-major axis of just over 19 AU.

It soon became apparent why Herschel had seen that it was a planetary body whilst others had not. Side by side comparisons with telescopes in use by others confirmed its far higher image quality – Herschel had proven to be a superb telescope maker! Uranus has a maximum angular size of 4.1 arcsec and, as the author has observed with an excellent telescope just a little smaller than Herschel's, it appears as a tiny greenish-blue disc. However, unless a telescope has well figured optics this disc would be very hard to distinguish from a star. The fact that the new planet had been charted, if not recognised as a planet, many times over the previous century allowed an accurate orbit to be computed and, later, this was to be a major factor in the discovery of Neptune.

Herschel quickly received acclaim and was made a Fellow of the Royal Society. The new planet became known as Uranus despite Herschel's wish for it to be named Georgium Sidus (George's star) after the King of England (a fellow Hanoverian). In 1782 Herschel demonstrated his telescope to the King and soon received a royal pension to allow him to devote himself exclusively to astronomy. Five years later, Caroline, who had become his observing assistant, also received a royal pension. To supplement their royal income Herschel made and sold telescopes including a 25ft (~7.6 m) long telescope that he made for the Madrid Observatory and for which he received £3000. In today's money this would be equivalent to at least £100 000!

Uranus revolves around the Sun once every 84 Earth years at an average distance from the Sun of roughly 3 billion km. The surface cloud layers are seen to rotate with a period of as little as 14 h, but this is due to high winds in the upper atmosphere and the nominal rotational period of Uranus is 17 h 14 min. Whilst for the majority of planets the rotation axis is roughly at right angles to the plane of the Solar System, Uranus has an axial tilt of 98°, so in effect 'rolls' around the Sun. Each pole gets around 42 years of continuous sunlight followed by 42 years of darkness. In contrast to the Earth, this makes the poles warmer than the equator.

Uranus is the least massive of the giant planets at 14.5 Earth masses and has the second lowest density ($1290 \, \mathrm{kg \, m^{-3}}$). It probably has a central rocky core of about 2 Earth masses above which is a mixture of various ices, such as water, ammonia, and methane, along with an outer gaseous layer made up of about 1 Earth mass of hydrogen and helium. As the ices make up a far greater proportion of its mass than gas, Uranus is often termed an ice giant rather than a gas giant.

## The rings of uranus

On March 10, 1977, observations were to be made of the occultation by Uranus of a star, SAO 158687, using a telescope mounted in the Kuiper Airborne Observatory. An occultation of a star occurs when the disc of the Moon or a planet passes in front of a star. An aircraft was used to observe the occultation partly to be above much of the Earth's atmosphere but also to bring the telescope to the narrow path across the Earth where the occultation could be observed. Just before the star's light is lost its light will have passed through the atmosphere of Uranus and comparing the spectra of the stars at this time with that prior to the occultation it is possible to learn about the planet's atmosphere.

The telescope was observing the star well before the expected time of occultation when the astronomers were somewhat perturbed as the star's light suddenly disappeared. The signal did return after a rather tense period but this was then followed by four partial losses of signal. Now reasonably confident that it was not their equipment that was faulty, the astronomers continued to observe the star following the occultation when the sequence was seen to repeat in the inverse order. The astronomers realised that the light from the star must have been eclipsed by material in five rings about Uranus with the outermost (called the epsilon ring) being the thickest. From the times of the ring's occultations they could calculate the diameters of the rings and found that the outermost was ~44 000 km from the centre of Uranus.

The rings were directly imaged when Voyager 2 passed Uranus in 1986 and more rings were discovered bringing the number up to 11 (Figure 3.21). In December 2005, the Hubble Space Telescope detected a pair of previously unknown rings which brought the number of rings to 13. These latter two lie considerably further out from Uranus and are thus termed its 'outer ring system'. The rings are not thought to be young and may have resulted as a moon suffered a high speed impact or came within the Roche limit of the planet when, as in the case of Saturn's rings, tidal forces would have split it into myriads of particles.

## 3.5.10    *Neptune*

| Mass | $1.0243 \times 10^{26}$ kg | 17.147 times that of Earth |
|------|------|------|
| Radius (equatorial) | 24 764 km | 3.883 times that of Earth |
| Radius (polar) | 24 341 km | 3.829 times that of Earth |
| Rotation period | 16 h 6 min 36 s | |
| Axial tilt | 28.32° | |
| Semi-major axis of orbit | 4 503 443 661 km | 30.1 AU |
| Orbital period | 60 190 days | 164.79 years |
| Average orbital speed | 5.43 km s$^{-1}$ | |
| Eccentricity | 0.0112 | |
| Orbital inclination | 1.77° | |

Neptune, at ~8th magnitude, can be seen in even a small telescope and had even been observed by Galileo (Figure 3.22). Whilst observing Jupiter on December 28, 1612 he recorded Neptune as an 8th magnitude star and a month later observed it close to a star on two successive nights. He noted that their separation had changed and could easily have reached the conclusion that this was because one was not a star but a planet! It was later observed by John Herschel, William's son, who also believed it to be a star.

The final discovery of Neptune is one of the most interesting stories in astronomy with its position being predicted independently by two mathematicians, John Couch Adams at Cambridge, and Urbain Le Verrier at the Paris Observatory. This resulted from a key fact about Uranus: its magnitude (which varies between 5.3 and 5.9) had meant that it had been observed many times prior to the realisation that it was a planet by William Herschel. John Flamstead had observed it several times from 1690 and allocated it the name 34 Tauri

**Figure 3.21**  An image of Uranus and its rings taken by the Voyager 2 spacecraft in 1986.
Image: Voyager 2, NASA.

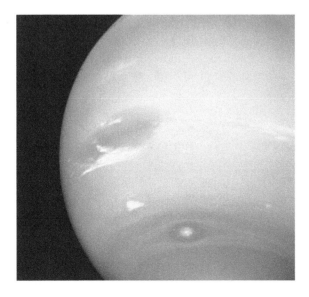

**Figure 3.22**  Neptune as imaged by Voyager 2 in 1989 showing the clouds, both light
and dark, in its atmosphere. Image: Voyager 2, NASA.

and it was recorded several times between then and its eventual discovery as a planet. These so-called 'ancient observations', when combined with more accurate observations made after its discovery as a planet, meant that an accurate orbit for Uranus could be immediately calculated. However, by 1821 it had become obvious that Uranus did not appear to be following its predicted obit and the thought arose that its orbit might be being perturbed by an, as yet undiscovered, planet that lay beyond Uranus.

John Couch Adams had studied at St John's College in Cambridge and graduated in 1843 as the 'senior wrangler', the best mathematician in his year. He was elected to a fellowship of his college and decided to devote his study to the resolution of the problem of the orbit of Uranus; believing it to be due to a more distant planet. He derived a solution in September 1845 and, it is thought, gave it to the Professor of Astronomy at Cambridge, James Challis. On September 21 he visited the home of the Astronomer Royal, George Airy but, failed to find him in. He tried again on October 21 but Airy was having dinner and would not see Adams who then left a manuscript giving his solution. Airy found this difficult to follow and sent a letter to Adam's requesting some clarification of the details, but it appears that Adams failed to reply – possibly because he had been upset at Airy's refusal to see him.

Meanwhile in Paris, Le Verrier, who had been asked to look at the problem by his director, had also derived a position for the planet. Perhaps because he was not well liked, there appears to have been no serious attempt to follow up on his prediction at the Paris Observatory and Le Verrier sent it to Airy. Airy realised the similarity of the predicted positions of the proposed planet and so, in July 1846, asked Challis (at Cambridge) to make a search for it. Unfortunately, Challis did not have a detailed star chart for the region where it was hoped that the new planet might be found so began to make one. If he then re-observed the positions of all the objects that he had charted, he would be able to spot the predicted planet by the fact that it would have moved slightly against the background stars.

On September 18 Le Verrier wrote to Johann Galle at the Berlin Observatory asking him to make a search. Galle received this letter on September 23 and his colleague, Heinrich d'Arrest, pointed out that they had a newly observed star chart for the appropriate region of the sky and thus they could easily spot the planet if it were present in the field. They made the observation that night and it took just 30 min to locate the planet just 1° away from Le Verrier's position!

In fact Challis had noted on September 29 that one of 300 stars that he charted that night had showed a disc – just as a planet would – but, being a cautious man, had waited for further observations to show its motion through the stars. He had not been able to make them before the announcement of the planet's discovery in the *Times* newspaper dated October 1, 1846.

William Lassell, who had made a fortune in the Brewery trade in Liverpool, had built a 24 in. telescope and immediately made observations of the new planet in the hope of finding any satellites. He started his observations on October 2 and on October 10 discovered Neptune's moon Triton.

The honour of Neptune's discovery is now shared by both Adams and Le Verrier. It is interesting to note that their predictions of Neptune's orbit were not that good and that for only about 10 years, from 1840 to 1850, did their predictions agree reasonably well with Neptune's actual position. More recently, the discovery of some papers that had been 'misappropriated' from the Royal Observatory at Greenwich cast some doubt on Adam's claim and perhaps he does not deserve equal credit with Le Verrier.

With hindsight, it is actually quite easy to predict where Neptune was to be found. The diagram in Figure 3.23 has been drawn in a rotating co-ordinate frame so that Neptune appears fixed in space. As Uranus is both orbiting the Sun more rapidly and its orbit has a shorter circumference, it will pass Neptune 'on the inside track'. You will see that, as Uranus nears Neptune, the gravitational force between them will tend to advance Uranus in its orbit, so it will appear ahead of where it would be expected to be (the solid disc, rather than the open circle). Once Uranus has passed beyond the position of Neptune, their gravitational attraction will slow Uranus down so that, eventually, it will have regained its expected position. One

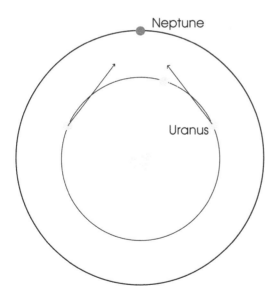

**Figure 3.23** The orbit of Uranus in a frame of reference in which Neptune is stationary.

can see that Neptune should lie beyond the position of Uranus when Uranus is furthest ahead of its predicted position.

Neptune is the fourth largest planet by diameter, and the third largest by mass, slightly more massive than its near-twin Uranus. Neptune's atmosphere is primarily composed of hydrogen and helium along with ~1% of methane which may help contribute to its vivid blue colour. Winds in its atmosphere can reach $2000 \, km \, h^{-1}$, the highest of any planet. As Voyager 2 passed Neptune it observed a Great Black Spot comparable with Jupiter's Great Red Spot. From the cloud tops at ~55 K to its core at 7000 K, Neptune has the greatest temperature range of any planet. Its sidereal rotation period is roughly 16.11 h long and it has a similar axial tilt to that of the Earth.

Neptune and Uranus are often considered ice giants, given their smaller size and greater percentages of ice in their composition relative to Jupiter and Saturn. Neptune's core is composed of rock and ice, having ~1 Earth mass. The mantle is made up of ~12 Earth masses largely made up of water, ammonia and methane. The atmosphere contains high clouds that cast shadows onto the blue coloured surface layers.

Like Uranus, Neptune also has a ring system. The rings, which have a reddish hue, may consist of ice particles coated with silicates or carbon-based material. In sequence measuring from the centre of Uranus are the broad, faint Galle Ring at 42 000 km, the Leverrier Ring at 53 000 km and the narrow Adams Ring at 63 000 km.

The largest of Neptune's 13 moons, and the only one massive enough to be spherical, is Triton which, unlike all other large planetary moons, has a retrograde orbit. This implies that it has been captured from what is called the Kuiper Belt, a region containing many small bodies beyond Neptune's orbit. It keeps one face towards Neptune and is slowly spiralling inwards where it will eventually be torn apart when it reaches the Roche limit giving Neptune a more extensive ring system.

## 3.5.11    *Pluto*

| Mass | $1.305 \times 10^{22}$ kg | 0.0021 that of Earth |
|---|---|---|
| Radius (mean) | 1.195 km | 0.19 that of Earth |
| Rotation period | −6.387 days (The minus sign signifies retrograde rotation.) | |
| Axial tilt | 119.59° | |
| Semi-major axis of orbit | 5 906 376 272 km | 39.48 AU |
| Orbital period | 90 613 days | 248.09 years |
| Eccentricity | 0.249 | |
| Orbital inclination | 17.142° | |

**Figure 3.24** Pluto with its three moons; Charon discovered in 1978 and Nix and Hydra discovered in 2005. Image: M. Mutchler (STScI), A. Stern (SwRI), and the HST Pluto Companion Search Team, ESA, NASA.

The discovery of Pluto followed on from that of Neptune (Figure 3.24). Neither the orbits of Uranus nor Neptune were well defined, but it was thought that there might be a more distant planet – called 'planet X' by Percival Lowell. In 1905 he had predicted its position and began a photographic search at the Flagstaff Observatory. Nothing was found and following more refined calculations he began a further search in 1914. Though Pluto did actually appear on some of the plates taken then it was not recognised.

Due to problems with Lowell's will following his death in 1916, the Observatory virtually ceased to function until 1929 when its then director, Vesto Melvin Slipher, began a new search. A young amateur astronomer, Clyde Tombaugh, had sent Slipher drawings that he had made of Jupiter using his Newtonian telescope in the hope of being offered a job (Figure 3.25). These impressed Slipher, and he employed Tombaugh to take images with the Observatory's 13 in. astrograph – essentially a wide field camera.

Two images taken some time apart were then compared in what is termed a 'blink comparator' in which the images are rapidly viewed in turn. Any object that has moved in the time between the two exposures would appear to jump in position whilst the stars would remain fixed. This allows planetary bodies to be rapidly located. The initial search for planet X was unsuccessful so Tombaugh began his own search.

From a pair of plates taken in January 1930, Tombaugh discovered a new planet on February 18, 1930. The motion of the planet was confirmed in follow

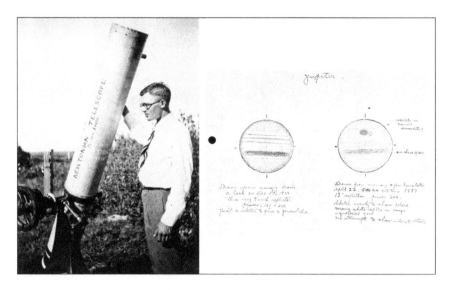

**Figure 3.25** Clyde Tombaugh with his home-built telescope along with drawings he made of the planet Jupiter. Image: New Mexico State University Library, Archives and Special Collections.

up observations and on March 13, 1930 its discovery was announced. Its name was suggested by Venetia Burney, the 11 year old daughter of an Oxford professor, when she was told of its discovery the next day. (Pluto was the Roman God of the underworld who was able to make himself invisible.) As the first two letters of the name Pluto were the initials of Percival Lowell, at whose observatory it had been discovered, this suggestion was eagerly accepted.

In the 1970s, James Christy had been making planetary observations at the Naval Observatory in Washington in order to refine their orbital parameters. He had been rejecting some images of Pluto as they had appeared somewhat elongated. This effect, as the author knows to his cost, usually appears when the telescope is not tracking correctly and images 'trail' so becoming elongated. However, he noted first that the star images in the photographic plate were perfect – indicating that the telescope *was* tracking correctly – and secondly, that the extension of Pluto's image appeared to move around Pluto as time went by. It transpired that Christy was observing the motion of a satellite, now called Charon, orbiting Pluto, and its discovery was announced on June 22, 1978. In 1990, the Hubble Space Telescope was able to image the two separate discs of Pluto and Charon. (Charon was the ferryman who carried the dead across the river Styx and had close ties to Pluto so was an ideal name. Its first four letters were also the same as those of Christy's wife Charlene!)

As described earlier in this chapter, if the period and semi-major axis of a satellite is known, it is possible to calculate the mass of the planet. For the first time it became possible to calculate the mass of Pluto which was found to be just 2% that of the Earth. This was far too low to have had the effects on the orbits of Uranus and Neptune by which its existence had been predicted.

In 2005, two further moons of Pluto were discovered, now called Nix and Hydra. Appropriately, Nix was the goddess of darkness and night and the mother of Charon. On January 18, 2006 a spacecraft called New Horizons left the Earth for a 9 year voyage to Pluto and beyond. Having flown over 3 billion miles, it will fly by Pluto and Charon around July 2015 and then hopefully visit some Kuiper Belt objects in an extended mission. New Horizons passed Jupiter on February 28, 2007 using the planet's gravitational pull to increase its speed to ~ 83 000 km h$^{-1}$ in what is called a slingshot manoeuvre and, in passing, made infrared images of Jupiter and its moons (Figure 3.26).

Pluto has a highly elliptical orbit inclined at over 17° to the ecliptic and has a mass and size both of which are far smaller that it was thought when first

**Figure 3.26** An infrared image of Jupiter and Io made by the New Horizons spacecraft as it passed Jupiter on its way to Pluto. Image: NASA, Johns Hopkins U. APL, SWRI.

discovered. There is no doubt that had it been discovered recently, it would never have been afforded the status of a planet. However, it has since become part of our culture and the author was somewhat saddened when it was demoted in 2006.

### 3.5.12    *Eris*

| Mass | $1.66 \times 10^{22}$ kg | |
|---|---|---|
| Radius (mean) | 1300 km ($+200/-100$ km) | |
| Rotation period | >8 h? | |
| Semi-major axis of orbit | $10.12 \times 10^9$ km | 67.67 AU |
| Orbital period | 203 600 days | 557 years |
| Eccentricity | 0.442 | |
| Orbital inclination | 44.2° | |

An object, 2003 UB$_{313}$, was discovered in images taken in 2003 in a survey being undertaken at the Mount Palomar Observatory searching for trans-Neptunian objects, that is, objects that lie beyond Neptune. The team, led by Mike Brown, had been very successful in the discovery of such objects including Quaoar, Orcus and Sedna. As the observations had been made in 2003, it is perhaps surprising that the discovery was not announced until 2005. The more distant an object, the less it will appear to move between images taken some time apart. In order to avoid false detections, the team had put a lower limit of 1.5 arcsec angular movement per hour. However, when Sedna was discovered moving at just 0.75 arcsec per hour, it was decided to reanalyse the collected pairs of images with a lower limit and it was then that the distant object, now at a distance of 96.7 AU from the Sun, was found. It has the highest albedo, 0.86, of any object in the Solar System save that of Saturn's moon Enceladus, perhaps as a result of the replenishing of surface ices.

Observations indicated that it was between 4% and 8% larger than Pluto and the discovery of a moon, now named Dysnomia, enabled its mass to be calculated. It is 27% more massive than Pluto. If Pluto were to be regarded as a planet so surely should 2003 UB$_{313}$. It was this that initiated the consideration of what should, and should not, be classed as a planet.

Formal naming of the object had to wait until it was determined whether it should be given the status of a planet. It was not until 2006, when along with Pluto and Ceres it was classified as a dwarf planet, that it gained its official name Eris, a Greek name personifying strife and discord – perhaps as a result of the upset it discovery had caused amongst the astronomical community (Figure 3.27)!

**Figure 3.27** An artist's impression of the dwarf planet Eris and its moon Dysnomia.
Image: Lombry Thierry.

## 3.6   Comets

Aristotle proposed the idea that comets were gaseous phenomena in the upper atmosphere that occasionally burst into flames (Figure 3.28). He depicted comets as 'stars with hair' and used the Greek word 'kometes' to refer to them from the root 'kome' meaning 'head of hair'. They were regarded as bad omens foretelling catastrophe or the deaths of kings. (Comet Halley is seen in the Bayeaux Tapestry in the sky above King Harold, who was killed by an arrow in 1066!)

**Figure 3.28** The comet Hale-Bopp seen close in the sky to the Andromeda Galaxy.
Image: J.C. Casado.

Tycho Brahe made careful observations of the comet of 1577 and, by measuring its position from well separated locations, was able to show that it lay at least four times further away than the Moon. In 1687, Isaac Newton was able to show that the path of a bright comet observed through the winter of 1680/1681 could be fitted to a parabolic orbit with the Sun at one focus. He had thus shown that comets were Solar System bodies orbiting the Sun.

### 3.6.1    *Halley's comet*

Edmund Halley calculated the orbits of 24 comets that had been observed between 1337 and 1698. He found that the comets that had been seen in 1531, 1607 and 1682 had very similar orbital elements and so believed that these were three 'apparitions' of the same comet. Halley could even account for the slight differences by taking into account the gravitational effects of Jupiter and Saturn. Its period was ~76 years so he predicted that it would return around the end of 1757. More accurate calculations by three French mathematicians indicated that it would, in fact, pass closest to the Sun in March 1759. It was first spotted by Johann Georg Palitzsch, a German farmer and amateur astronomer on Christmas Day 1758. Halley, who had died in 1742, did not live to see the comet's return but it was named after him with the designation 1P/Halley – the '1P' indicating that it was the first known periodic comet.

Dependent on the Earth's orbital position as Halley nears the Sun, it can appear both bright and spectacular as it did in 1066 and 1910 or, as in its last apparition in 1985/1986 when far from the Earth, barely visible to the unaided eye. The comet was first photographed in 1910 when it made a very close approach to the Earth – which even passed through its tail causing some alarm! It is thought that the artist Giotto di Bondone observed the comet in 1301, and used it to depict the Star of Bethlehem in his 1305 fresco, 'The Adoration of the Magi'(Figure 3.29). As a result, the spacecraft sent to fly-by Comet Halley in 1986 was called Giotto. Halley's Comet will next appear in our skies in 2061.

A comet is now classified as a small Solar System body that orbits the Sun and which, when close to the Sun, exhibits a visible coma (an extended atmosphere) and sometimes a tail. The 'nucleus' of a comet is typically of order 10 km in size and is composed of rock and dust bound together by ice. The term 'dirty snowball'" that is sometimes used is thus quite apt. Those that are termed 'long-period' comets are debris left over from the condensation of the solar nebula and come from the outermost regions of the Solar System, up to a light year distant from the Sun, in what is usually termed the Oort cloud.

The cloud is thought to contain of the order of a trillion comets and is named after the Dutch Astronomer, Jan Hendrik Oort, who popularized the idea in 1952.

**Figure 3.29** Giotto de Bondone's 'Adoration of the Magi'. Giotto had seen Halley's Comet in 1301 and depicted it as the Star of Bethlehem in his fresco painted in 1305.

However, the concept was first proposed by an Estonian Astronomer, Ernst Opik, in 1932 so it is alternatively, and more correctly, called the Opik–Oort cloud. Such comets will normally only ever be seen once, but occasionally their orbit will be sufficiently perturbed by Jupiter or Saturn to be 'captured' within the inner Solar System with a relatively short period. It will then become known as a 'short-period' comet which, by definition, has a period of less than 200 years. However, the majority of short-period comets are thought to originate in the Kuiper Belt, which lies beyond the orbit of Neptune. Of the ~3000 comets known by the end of 2007, several hundred have short periods. On average, about one comet per year will reach unaided-eye visibility, but only about one in ten of these will become easily visible.

## 3.6.2    *Cometary nuclei*

The nuclei of comets range from ~0.5 to 50 km in size and have a very rich composition; primarily made of rocks, dust and water ice along with frozen gases, such as carbon dioxide, carbon monoxide, methane and ammonia. They also contain many organic compounds such as methanol, formaldehyde, ethanol and ethane and possibly even more complex molecules, such as amino acids. They are far too small to become spherical through the force of gravity and are thus irregularly shaped. Comets cannot normally be seen in the outer Solar System due to their small size and low albedo, and only become visible as they approach the Sun whose heat radiation vaporizes the water ice and other volatile materials. This releases the dust and a giant tenuous cloud, called the coma, of dust and gas forms around the nucleus reflecting the light from the Sun and so making the comet visible.

The Sun's radiation pressure and solar wind then come into play causing two tails to form. The gases, in the form of ions, are most strongly affected by the solar wind and always appear to point directly away from the Sun – as first shown by the German astronomer Peter Apian in 1531. The ion tail often appears bluish in colour due to emission from $(CN)_2$, cyanogen. The dust tail is yellowish in colour and tends to lie along the orbit of the comet so often appears curved. The coma may exceed the Sun in size, and ion tails have been known to extend over 1 AU in length, and so are the largest objects in the Solar System. The size of the coma and tail increases as the comet nears the Sun and the comet will tend to be most apparent immediately after it has passed the Sun. We thus tend to observe them in the period before dawn or after dusk. With each passage around the Sun, the comet loses material and will eventually disintegrate into a trail of dust or become an inert, asteroid-like body of fractured rock.

The fact that comets may contain significant amounts of organic compounds is indicated by their very low albedo. The Giotto space probe showed that Comet Halley's nucleus reflected only ~4% of the incident sunlight whilst Deep Space 1 showed that Comet Borrelly only reflected ~2.6%. It is thought that the dark surface is made up of complex, tar-like organic compounds and hence that cometary impacts on Earth may have brought much organic material to our surface. It is also thought that such impacts may have brought much of the water found on the planet and they most certainly have had an effect on the direction of evolution here on Earth. We have a lot to thank them for – apart from providing us with some of the most beautiful sights in the heavens!

## 3.7   Questions

1.   An eclipse of the Sun is to occur on the day of the vernal equinox and when the Earth is at a distance of 150 million km from the Sun. The eclipse track passes over the equator at 12:00 Local Solar Time. Show that, if the Moon is at its furthest point from the Earth (at apogee) that day, the eclipse as observed from the equator will be an annular, not a total, eclipse.

   Show also that if the Moon were to be closest to the Earth that day (at perigee) there would be a total eclipse.

(The equatorial radius of the Earth is 6378 km. The semi-major axis of the Moon's orbit is 384 401 km and the eccentricity of its orbit is 0.056. Its diameter is 3474.8 km. The diameter of the Sun is 1.39 million km.)

2.  A rapidly rotating asteroid lies four times further from the Sun than the Earth. Assuming the asteroid acts like a black body and is roughly spherical, estimate its surface temperature. (The solar constant at the Earth's distance is 1370 W m$^{-2}$ and Stefan's constant is 5.7 × 10$^{-8}$ W m$^{-2}$ K$^{-4}$.)

3.  Show that the surface temperature, $T$, of a rotating spherical body at a distance $R$ from the Sun falls as the inverse square root of $R$.

   Knowing that a spacecraft in Earth orbit prior to orbital insertion towards the Sun reached a stable temperature of 260 K, find the constant of proportionality in the relationship between $R$ and $T$, where $T$ is in kelvin and $R$ is in astronomical units.

   How close (in AU) could this spacecraft be allowed to approach the Sun if it could withstand a surface temperature of 1000 K?

4.  Whilst Venus was making its closest approach to Earth, a Doppler radar, transmitting at a frequency of 1420 MHz, received an echo from Venus which was 'spread' over a bandwidth of 17.1 Hz. Calculate the speed at which the receding limb is travelling away from us with respect to the centre of Venus and hence estimate, in Earth days, how long Venus takes to make one rotation about its axis.

   (Assume the velocity of light is 3 × 10⁵ km s$^{-1}$. The non-relativistic Doppler formula is given by $\Delta f/f = \Delta v/c$. Venus has a diameter of 12 104 km.)

# Chapter 4

# Extra-solar Planets

This is one of the most exciting areas of research being undertaken at the moment with the discovery of new planets being announced on a monthly basis. This chapter will describe the techniques that are being used to discover them and then discuss their properties. Perhaps a word of warning might be in order. An obvious quest is to find planetary systems like our own which could, perhaps, contain planets that might harbour life. So far, to many astronomers' surprise, the vast majority of solar systems found have been very unlike our own which might lead one to the conclusion that solar systems like ours are very rare. The author has even heard this point of view put forward by an eminent astro-biologist. At this time, one should not draw this conclusion. For reasons which shortly become apparent, the techniques largely used to date would have found it very difficult to detect the planets of our own Solar System so it should not be surprising that we have so far failed to find many other similar solar systems. As new techniques are used, this situation will improve, but it will be many years before we have any real idea how often solar systems like our own have arisen in the galaxy. The story of the discovery of the first planet to orbit a sun-like star is very interesting in its own right, but, in order to appreciate its nuances, we need first to understand how this, along with the great majority of planets so far detected, has been discovered.

## 4.1 The radial velocity (Doppler wobble) method of planetary detection

Our own Solar System gives us a good insight into this method and its strengths and weaknesses. Astronomers often use, as in this book, the phrase 'the planets orbit the Sun'. This is not quite true. Imagine a scale model of the Solar System with Sun and planets having appropriate masses and positions in their orbits from the Sun. All the objects are mounted on a flat, weightless, sheet of supporting material. By trial and error, one could find a point where the model could be balanced on just one pin. This point is the centre of gravity of the Solar System

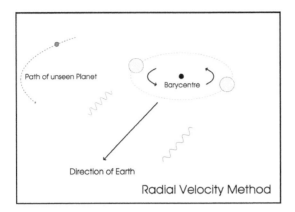

**Figure 4.1** The change in wavelength of a spectral line as the star orbits the barycentre of its solar system.

model. The centre of gravity of the Solar System is called its barycentre, and *both* the Sun and planets rotate about this position in space (Figure 4.1).

As Jupiter is more massive than all the other planets combined, its mass and position have a major effect on the position of the barycentre, which will thus lie a distance from the centre of the Sun in the approximate direction of Jupiter. How far might this be?

Ignoring all other planets, let us assume that the barycentre lies $R$ km from the centre of the Sun. The mass of the Sun and Jupiter must 'balance' about this point so, given $R$ and the semi-major axis of Jupiter's orbit in km:

$$M_{Sun} \times R = M_{Jupiter} \times (777\ 547\ 199 - R)$$

The Sun, with a mass of $2 \times 10^{30}$ kg, is approximately 1000 times more massive than Jupiter. Hence:

$$1000\,R + R = 777\ 547\ 199$$

So,
$$R = 777\ 547\ 199/1001$$
$$= 776\ 770\,\text{km}$$

The equatorial radius of the Sun is 695 500 km so, if Jupiter is its only planet, the barycentre of the Solar System would actually lie outside the Sun. When all of the major planets lie on one side of the Sun, as happened in the 1980s – allowing the Voyager spacecraft missions to the outer planets – the barycentre is further from the Sun's centre and when Jupiter is on the opposite side to the other planets it is nearer the Sun's centre. On average, the barycentre is at a distance of ~1.25 solar radii from the Sun's centre, varying between extremes of ~0.3 and 2 solar radii.

Suppose that we observed the Solar System from a point at a great distance in the plane of the Solar System. We could not see the planets – their reflected light would be swamped by the light from the Sun – but, at least in principle, we could detect their presence. Due to the Sun's motion around the barycentre of the Solar System, it would at times be moving towards us and at other times moving away from us. If we could precisely measure the position of the spectral lines in the solar spectrum we could measure the changing Doppler shift and convert that into a velocity of approach or recession. The Solar System as a whole might, of course, be moving either away or towards us so we would see a cyclical change in velocity about a mean value.

Again, for the sake of simplicity, let us assume that our Solar System has only one planet (Jupiter).

The Sun would be seen to rotate around the barycentre once every 11.86 years, the period of Jupiter's orbit. Given our calculation of the position of the barycentre, we can thus calculate the speed of the Sun in its orbit about the barycentre. The circumference of the Sun's orbit about the barycentre is:

$$2 \times \pi \times 7.77 \times 10^5 \, \text{km} = 4.9 \times 10^6 \, \text{km} = 4.9 \times 10^9 \, \text{m}.$$

So, as 11.86 years is $3.74 \times 10^8$ s, the orbital speed is:

$$4.9 \times 10^9 / 3.74 \times 10^8 = 13 \, \text{m s}^{-1}.$$

This would mean that the difference between the maximum and minimum velocities would be ~26 m s$^{-1}$.

The current precision in Doppler measurements is of order 2–3 m s$^{-1}$, but the hope is that, in time, this might improve to ~0.5 m s$^{-1}$. Very high resolution spectrometers are used to observe the light from the star whose light is first passed through a cell of gas to provide reference spectral lines to allow the Doppler shift to be measured.

The measurement accuracy of this method would thus be sufficient to detect the presence of Jupiter in orbit around the Sun. However, in order to be reasonably sure about any periodicity in the Sun's motion one would need to observe for at least half a period and preferably one full period. So observations have to be made on a time scale of many years in order to detect planets far from their sun. This is the major reason why few planets in large orbits have yet been detected – the observations have simply not been in progress for a sufficiently long time.

There is one other limitation that you might have realised: should we observe a distant solar system from directly above or directly below, then we would see no Doppler wobble and hence could not detect any of its planets. Unless we have additional information that can tell us the orientation of the orbital plane of a distant solar system, we can only measure the minimum mass of a planet, not its actual mass. If, for example, we later observed such a planet transit across the face of its sun then we would know that the plane of its solar system included the Earth so that the derived mass is the actual mass of the planet, rather than a lower limit.

A single planet in a circular orbit will give rise to a Doppler curve which is a simple sine wave. If the orbit of the planet is elliptical, a more complex, but regularly repeating Doppler curve results. In the case of a family of planets, the Doppler curve is complex and will not repeat except on very long timescales. It can, however, still be analysed to identify the individual planets in the system.

In a manner similar to the way in which we calculated the orbital motion of the Sun due to Jupiter, one could calculate the Sun's orbital velocity due to the Earth. This is $0.1\,\mathrm{m\,s^{-1}}$, well below the current and predicted future sensitivity of the radial velocity method so other methods are required for the detection of Earth-like planets. As other techniques (discussed below) come to fruition and longer periods of observation are analysed by the radial velocity method, solar systems like our own are beginning to be found but, as yet, we cannot say how common they are.

### 4.1.1    *Pulsar planets*

There is one case where Doppler measurements can be made to extreme accuracy. This is when the central object is the remnant of a giant star called a neutron star. These will be discussed in detail in Chapter 7, but all we need to know at this time is that some of these stars (called pulsars) emit regular, very precisely timed pulses and so Doppler shifts can be measured to exceedingly high precision. This, in principle, would easily allow Earth-mass planets to be discovered. However, it is thought that planets would not often survive the massive nuclear explosion – called a supernova – when the pulsar is formed, so such planets are likely to be very rare. However, in 2006 observations made using the Spitzer Space Telescope showed that the pulsar 4U 0142+61 had a circumstellar disc. The disc is thought to have formed from metal-rich debris left over from the supernova explosion that had given rise to the pulsar and is similar to those seen around Sun-like stars, suggesting that planets might be able to form within it. Pulsars emit vast amounts of electromagnetic radiation so such planets would be totally incapable of supporting any form of life!

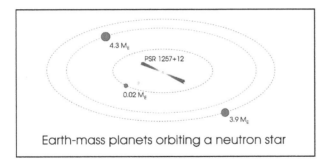

Earth-mass planets orbiting a neutron star

**Figure 4.2** Planets orbiting the pulsar PSR 1257+12 at distances of 0.19, 0.36 and 0.46 AU.

This is perhaps how the planets were formed in orbit around the pulsar B1257+12. Two planets were initially discovered in 1992 by Aleksander Wolszczan and Dale Frail. Their orbits would both fit within the orbit of Mercury and they had masses of 4.3 and 3.9 Earth masses (Figure 4.2). Two further planets with masses of just 0.004 and 0.02 Earth masses have since been found in the system and, more recently, a single 2.7 Jupiter-mass planet has been found in orbit around the pulsar B1620–26. It is possible that three other pulsars have one or more planets in orbit around them, but these have yet to be confirmed.

### 4.1.2    *The discovery of the first planet around a sun-like star*

In 1988, Canadian astronomers, Bruce Campbell, G. A. H. Walker, and S. Yang, suggested from Doppler measurements that the star Gamma Cephei might have a planet in orbit about it. The observations were right at the limit of their instruments capabilities and were largely dismissed by the astronomical community. Finally, in 2003, its existence was confirmed but, unfortunately, this was many years after the first confirmed discovery of a planet around a main sequence star.

Two American astronomers, Paul Butler and Geoffrey Marcy, were the first to make a serious hunt for extra-solar planets. They began observations in 1987 but, assuming that other planetary systems were similar to our own, did not expect that any planets could be extracted from the data for several years. They would have thus been somewhat shocked when the discovery of a planet orbiting a star called 51 Pegasi was announced by Michael Mayor and Didier Queloz on October 6 1995. The star 51 Pegasi, or 51 Peg for short, lies just to the right of the square of Pegasus and is a Type G5 star, a little cooler than our Sun, with a mass of 1.06 solar masses. Meyer and Queloz were studying the pulsations of stars, which also causes a Doppler shift in the spectral lines as the star 'breathes' in and out. With

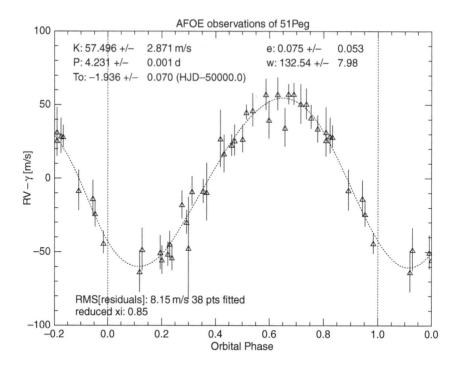

**Figure 4.3** Radial velocity measurements of 51 Pegasi made by Korzennik and Contos using the advanced fibreoptic echelle spectrometer on the 1.5 m telescope at the Whipple Observatory near Tucson, Arizona.

a sensitivity of only $15\,\mathrm{m\,s^{-1}}$ they had not really expected to discover planets but, much to their surprise, they found a periodicity in the motion of the star 51 Peg having a period of 4.23 days and velocity amplitude of $57\,\mathrm{m\,s^{-1}}$. The plot is very close to a sinusoid showing that the orbit is very nearly circular (Figure 4.3).

Let us calculate the mass and orbital radius of the planet which is called 51 Pegasi b (Figure 4.4). From the velocity of the star and the period of the orbit we can first calculate the circumference and hence the radius of the star's motion: 4.23 days is 365 472 s, so the circumference is $57 \times 365\,472\,\mathrm{m} = 20\,831\,904\,\mathrm{m}$, giving a radius of 3 315 500 km. This is thus the distance from the centre of the star to the barycentre of the system.

You may remember that we were able to calculate the mass of the Sun given the orbital period and the distance of the Earth from the Sun:

$$M = 4\pi^2 a^3 / G P^2$$

**Figure 4.4** Artist's impression of 51 Pegasi b orbiting its sun. Image: Wikipedia Commons.

(Note: this slightly simplified equation assumes the mass of the planet is far less than that of the star.)

As we know the mass of the star 51 Peg ($M = 1.06 \times 2 \times 10^{30}$ kg), and the period $P$, we can solve this equation for $a$, the distance of the planet from the star, given the universal constant of gravitation, $G$.

$$a = (GM\,P^2/4\pi^2)^{1/3}$$

Substituting:

$$= [6.67 \times 10^{-11} \times 2.12 \times 10^{30} \times (365\ 472)^2/4\pi^2]^{1/3}$$
$$= 7.82 \times 10^9\,\text{m}$$

One AU is $1.496 \times 10^{11}$ m so the planet lies at a distance of 0.052 AU from 51 Peg – this is well within the distance of 0.39 AU at which Mercury orbits our Sun and only about 10 times the radius of the star.

We can now find the mass of the planet by balancing about the barycentre of the system, which we have calculated lies at a distance of 3 315 500 km from the centre of the star. We have:

$$M_{\text{planet}} \times 7.81 \times 10^9 = M_{\text{star}} \times 3\ 315500$$

Giving the mass of the planet:

$$M_{\text{planet}} = 2.12 \times 10^{30} \times (3\ 315\ 500/7.81 \times 10^{9})\ \text{kg}$$
$$= 9 \times 10^{26}\,\text{kg}$$

Jupiter has a mass of $1.9 \times 10^{27}$, so the planet has a calculated mass 0.47 that of Jupiter.

However, this would only be the mass of the planet if the plane of its orbit included the Earth and will thus be the planet's minimum mass. One can show that, for random orientations, the mass of a planet will on average be about twice the minimum mass, so the planet in orbit around 51 Peg is likely to be very similar in mass to Jupiter.

When Butler and Marcy learnt about this discovery, they realized that not only could they confirm its presence from several years observations of 51 Peg in their database – which they did just 6 days later – but that if other massive planets with short periods existed around the stars that they had been observing, they should be able to rapidly find these as well. This hope was borne out and, to date, they have been the world's most prolific planet hunters.

No one had expected that a gas giant would be found so near its star, but many of the planets first discovered were similar in size and separation from their sun. It is not thought that giant planets can form so close to a star, so at some time in their early history it is assumed that they must have migrated inwards through the solar system. In doing so, they would very likely eject smaller (terrestrial type) planets that had formed nearer the star from the solar system and consequently these solar systems are thought unlikely to harbour life.

The Doppler wobble method has proven to be highly successful, but even with a hoped-for velocity precision of $0.5$–$1\,\text{m s}^{-1}$ at best, the method will never be able to detect Earth-mass planets, no matter how close they are to their sun.

## 4.2    Planetary transits

As the number of known close orbiting gas giants increases, there becomes a reasonable chance that the plane of some of their orbits will include the Earth and so, once each orbit, the planet might occult the star, giving a measurable drop in its brightness (Figure 4.5).

Let us estimate the magnitude drop if a Jupiter sized planet occulted our Sun as seen from a great distance. The Sun has a diameter which is ~10 times that of Jupiter, so that its cross-sectional area will be ~100 times that of Jupiter. When Jupiter occulted the Sun, the effective area will drop from 100 to 99 – a ratio of

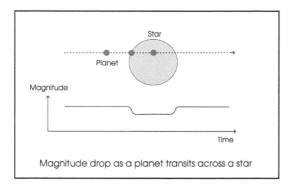

Magnitude drop as a planet transits across a star

**Figure 4.5** The effect of a planetary transit on the brightness of a star.

0.99 – and give a drop in brightness of 1%. This corresponds to a magnitude drop given by:

$$\Delta m = 2.5 \log_{10}(0.99)$$
$$= -0.011 \text{ magnitudes.}$$

With care, such accuracy in measurement is achievable and on November 5, 1999 two teams detected the transit of a planet, previously discovered by the radial velocity method, in orbit around the star HD 209469. During the transit, the brightness of the star dropped by 1.7%.

In 2002, a planet OGLE-TR-56B was discovered by the transit method and later confirmed using the radial velocity method. Then, in 2006, the Hubble Space Telescope made a survey of 180 000 stars up to 26 000 light-years away towards the central bulge of our Galaxy. The survey discovered 16 candidate extra-solar planets of which three have since been confirmed. Such confirmation is required as the technique has a high rate of false detections. If all 16 were confirmed, it would imply that there would be of order 6000 million Jupiter sized planets in the galaxy. Five of the newly discovered planets were found to orbit their sun with periods of <1 day. The candidate with the shortest period – just 10 h – is only 1.2 million km from its relatively small, red dwarf sun and has an estimated surface temperature of 1400 K! It must be at least 1.6 times the mass of Jupiter in order to prevent the tidal forces from the star splitting the planet apart (Figure 4.6).

Apart from the high rate of false detections the transit method has the problem that transits can only be observed when the planet's orbit is nearly edge on. About 10% of planets in close orbits would show transits, but the fraction is far smaller for planets with large orbits as the alignment has to be more precise – only ~0.5%

**Figure 4.6** An artist's impression of a transiting exoplanet. Image: ESA – C. Carreau.

of Earth-like planets in orbit around stars similar to our Sun would cause transits. Two space missions called Kepler and COROT will have very large field of views, enabling them to continuously monitor many stars. The 95 Megapixel CCD array used with the 0.95 m Kepler telescope will monitor more than 100 000 stars with very high precision during an initial 3.5 year observing period. It is hoped that Kepler and COROT will, for the first time, enable the detection of a significant number of Earth sized planets.

The transit method does have two significant advantages. The first is that, as a planet will take some time to fully cover its star, the size of the planet can be determined from the light curve. When combined with the planet's mass, determined by the radial velocity method, the density of the planet can be determined and so we can learn about its physical structure.

The second advantage is that it is possible to study the atmosphere of a planet. When the planet transits the star, light from the star passes through the atmosphere of the planet. By carefully studying the star's spectrum during the transit, absorption lines will appear that relate to elements in the planetary atmosphere.

The extra-solar planet HD 209458b, provisionally nicknamed Osiris, was the first planet observed transiting its sun (Figure 4.7). Observations by the Hubble Space Telescope first discovered a tail of evaporating hydrogen which may, in time, completely strip the planet of gas leaving a 'dead' rocky core. More recent Hubble Space Telescope observations have shown that the planet is surrounded

**Figure 4.7** An artist's impression of the planet HD 209458b showing an extended envelope of carbon and oxygen and tail of evaporating hydrogen. Image: ESA and Alfred Vidal-Madjar (Institut d'Astrophysique de Paris, CNRS, France).

by an extended envelope of oxygen and carbon believed to be in the shape of a rugby ball. These heavier atoms are caught up in the flow of the escaping atmospheric atomic hydrogen and rise from the lower atmosphere rather like dust in a whirlwind.

## 4.3    Gravitational microlensing

The detection of Earth-mass planets by the transit method still (in 2007) lies in the future, but there is a third method which has the potential to achieve this now – and has already detected a 5 Earth-mass planet. In Chapter 2 we saw how Einstein's General Theory of Relativity was proven by the observed movement of star's positions due to the curvature of space close to the Sun. This effect gives rise to what is called gravitational lensing, specifically gravitational microlensing as the effects are on a very small scale. In the same way that a convex lens can concentrate the light from a distant object into the eye and so make it appear brighter, if a distant star passes behind one of intermediate distance, the brightness of the distant star will undergo a temporary increase which can last for many days. The peak brightness can be up to 10 times (2.5 magnitudes) that normally observed. More than a thousand such events had been observed by the end of 2007.

If the lensing star has a planet in orbit around it, then that planet can produce its own microlensing event, and thus provide a way of detecting its presence. For this to be observed, a highly improbable alignment is required so that a very large number of distant stars must be continuously monitored in order to detect planetary microlensing events. Observations are usually performed using networks of robotic telescopes (such as those forming the OGLE collaboration) which continuously monitor millions of stars towards the centre of the galaxy in order to provide a large number of background stars.

If one of the telescopes finds that the brightness of a star is increasing, then the whole network, spaced around the world for continuous observation, will provide unbroken monitoring. The presence of a planet is shown by a very short additional brightening appearing as a spike on the flanks of the main brightness curve.

On January 25, 2006, the discovery of OGLE-2005-BLG-390Lb was announced (Figure 4.8). This planet is estimated to have a mass of ~5.5 Earth masses and orbits a red dwarf star which is around 21500 light years from Earth, towards the centre of the Milky Way Galaxy (Figure 4.9). The planet lies at a distance of 2.6 AU from its sun. At the time of its discovery, this planet had the lowest mass of any known extra-solar planet orbiting a main sequence star. This record may still hold unless the mass of Gliese 581c, discovered in April 2007 by the radial velocity method, is found to have a mass very close to its minimum mass of 5 Earth masses. By the end of 2007, four extra-solar planets had been discovered using the microlensing technique.

A disadvantage of the method is that the chance alignment that allowed the lensing event that led to the planet's detection is highly unlikely to be ever repeated. Also, the detected planets will tend to be many thousands of light years

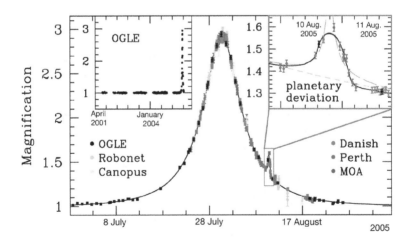

Light Curve of OGLE-2005-BLG-390

**Figure 4.8** Observations by the OGLE consortium showing the microlensing caused by a planet of 5.5 Earth masses. Image: ESO press release March 2006.

**Figure 4.9** Artist's impression of the 5.5 Earth-mass planet which circles its red dwarf star with a period of 10 years. The planet has surface temperature of ~220°C below zero. It is likely to have a thin atmosphere with a rocky core buried beneath a frozen icecap. Image: ESO press release March 2006.

away, so making any follow-up observations by other methods virtually impossible. However, if enough background stars can be observed over long periods of time the method should finally enable us to estimate how common Earth-like planets are in the galaxy.

In early 2008, the gravitational microlensing method detected two gas giant planets, similar to Jupiter and Saturn, orbiting a star 5000 light years away in a planetary system with striking similarities to our own Solar System. The discovery suggests that giant planets do not live alone but are more likely to be found in family groups. The mass of the nearer planet is 0.71 times that of Jupiter and it lies 2.3 times as far from its host star as the Earth is from the Sun. The second planet is less massive; 0.27 times the mass of Jupiter, and twice as far away from the host star.

Despite their host star only being half as massive as the Sun the planetary system otherwise bears a remarkable similarity to our Solar System. Both the ratio of the masses of the two giant planets (close to 3:1) and the ratio of their distances from the host star (1:2) are remarkably similar to those of Jupiter and Saturn. The ratio between the orbital periods of 5 years and 14 years, respectively, also closely resembles that between Jupiter and Saturn (2:5). The newly discovered system resembles our own Solar System more closely than any previously observed. Whilst there are more than 250 planets now known to be orbiting other stars, there are only about 25 solar systems known to have multiple planets; this number will surely rise as smaller planets fall within our detection methods.

## 4.4    Astrometry

The science of accurate positional measurement, called astrometry, is the oldest method that has been employed in the search for planets and in the 1950s and 1960s the discovery of several planets was claimed using this method. Sadly, none has since been confirmed and, by the end of 2007 no planets had been discovered by this method, though the Hubble Space Telescope has confirmed the existence of a planet in orbit around Gliese 876.

The method consists of observing, with great precision, how a star's position varies over time. All star systems are moving around the galaxy (our Sun in ~230 million years) and over short periods of time will move in essentially straight lines. However, if the star is in a binary or planetary system, the point that moves in a straight line is not the centre of the star but the barycentre of the system. So, as the star moves in a tiny circular or elliptical orbit about the barycenter, it will follow a wiggly path across the sky. As yet, except in the case of Gliese 876, it has not been possible to produce observations of sufficient accuracy to allow the presence of a planet to be detected. A simple calculation might indicate why this is so. Consider the star 51 Pegasi. It lies at a distance of 3 315 500 km from the barycentre of its planetary system. Assuming that we are at right angles to the plane of its planet – which is the most favourable condition – then we could calculate the amplitude of the angular oscillation that would be observed as 51 Pegasi moves across the heavens. (Note that if the plane of 51 Pegasi's planet was at right angles to us, the radial velocity method could not have detected it!)

Let us take a distance for the star system of 30 light years which is ~$3 \times 10^{14}$ km. The angular motion in radians is thus ~$3 \times 10^6 / 3 \times 10^{14}$, which is $1 \times 10^{-8}$ rad or 0.002 arcsec. The angular resolution of the Hubble Space Telescope is 0.05 arcsec but a very clever technique has enabled it to make measurements with a positional accuracy of 0.0005 arcsec and so it would be able to detect the planet around 51 Pegasi. As mentioned earlier, the Hubble Space Telescope has been able to detect the planet in orbit around Gliese 876, which lies at a distance of 15.6 light years and was first detected with the radial velocity method in 1998. The radial velocity measurements combined with 2 years of the Hubble Space Telescope astrometric measurements allow the orientation of the plane of the planetary orbits to be determined so that the actual, not just the minimum, mass of the planets in that system has been found.

Future spacecraft, such as NASA's Space Interferometry Mission (SIM), should finally attain a measurement precision of 1 μarcsec or 0.000001 arcsec. The mission would thus have the capability to detect planets at considerably greater distance than 30 light years.

The astrometric method nicely complements the radial velocity and transit methods in that it is more sensitive to planets at larger orbital distances. However, such planets will have long orbital periods so observations over many years would be required for their detection.

## Discovery space

Figure 4.10 is complex but very instructive. It shows the range of planets, in both mass and separation from their star, which could be detected by the four methods that have been discussed above. The lower horizontal axis is the distance (in AU) from the star with, above, the upper horizontal axis giving the corresponding orbital period of a planet around a 1 solar mass star. The vertical axes give the mass of a planet in both Jupiter and Earth masses. The plot shows the position, given by their mass and period, of the planets of our Solar System along with a number of extra-solar planets to give an indication of the parts of the discovery space in which the planets have been found by the differing methods.

In blue are the planets discovered by the radial velocity method. They lie to the upper left of the diagram in that they are both massive and tend to lie close to their star. The two blue lines separated by a pale blue shaded area show the discovery space with Doppler precisions of $3\,\mathrm{m\,s^{-1}}$, relating to the year 2006, and $1\,\mathrm{m\,s^{-1}}$, the expected precision that will be achieved by the year 2010. You will see that planets like Jupiter in mass and separation from their star could already have been detected by this method and that those like Saturn could follow. The fairly sharp cutoff at a period of 10 years is simply due to the fact that we need to observe a star for at least one orbital period to be sure of its presence and observations had only been taking place for ~10 years at the time this plot was made.

In red are the planets discovered by Earth based transit observations. Their discovery space is relatively small, but will greatly increase when the two space missions COROT, launched in December 2006, and Kepler, due for launch in February 2009, complete their missions. The initial planetary detections need extensive follow-up observations to confirm their presence much of which is being done from ground based telescopes. By the end of 2007, many candidate planets found in the first year of COROT transit observations were being followed up from the ground and the discovery of two planets had been announced. They are shown in green on the plot.

By 2004 one planet, shown in yellow, had been discovered by gravitational microlensing. Three further planets, shown in orange, had been discovered by the end of 2007. As mentioned above, one of these has the lowest *known* mass of any extra-solar planet, 5.5 Earth masses.

As yet, no planets have been detected by the astrometric method. Two lines in the upper part of the diagram show the discovery space of two ground based surveys. The upper blue line relates to a Palomar survey using the 5-m Hale

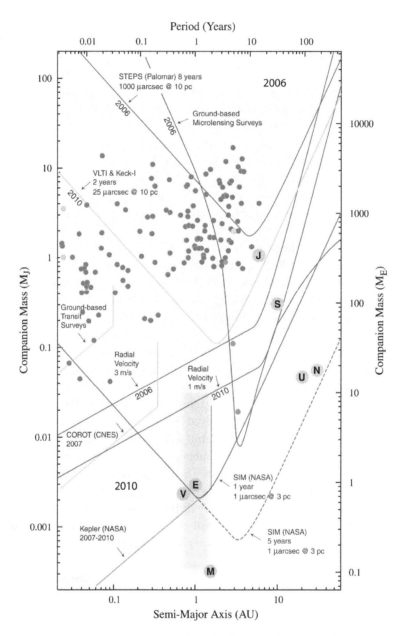

**Figure 4.10** The discovery space of extra-solar planets as described in the text.
Image: Wikipedia Commons.

telescope, whilst the lower green line relates to the surveys to be undertaken by the VLT and Keck telescopes. The solid and dotted red lines show the discovery space that could be achieved by NASA's SIM spacecraft after 1 and 5 years from launch, respectively. Sadly, development work on the project has slowed due to budget cuts and its launch has been delayed indefinitely. As the mission would significantly increase the discovery space for planets of similar mass and separation from their star as our Earth, it is hoped that its launch will not be delayed for too long.

## 4.6    Selection effects and the likelihood of finding solar systems like ours

By the end of 2007, the vast majority of the extra-solar planets found have high masses – many considerably more massive than Jupiter. Only six have masses less than 10 times that of the Earth. However, as all the detection methods described above are far more likely to discover massive planets, this is largely an observational selection effect. It is encouraging that despite the great difficulty of detecting them, astronomers have found several planets only a few times more massive than Earth. This implies that such planets are actually quite common. The other selection effect is that there has only just been enough time since radial velocity observations began to detect Jupiter-mass planets at similar distances from their sun. It may very well be that solar systems like ours *are* quite common despite what might appear at first sight.

## 4.7    Questions

1.  The radial velocity of a 1 solar mass star is seen to vary sinusoidally by $+/-30\,\mathrm{m\,s^{-1}}$ with a period of 1460 days. Assuming that the Earth lay in the plane of the orbit of the object causing this stellar motion, calculate its distance from the star in astronomical units (AU) and its mass in Jovian masses. (Assume that the simple form of Kepler's Third Law is valid, where $P^2 = R^3$ when $P$ is in years and $R$ is in AU. $1\,\mathrm{AU} = 1.49 \times 10^{11}\,\mathrm{m}$, and Jupiter is 0.001 solar masses.)
2.  The current precision of the Doppler shifts in the absorption lines in stellar spectra is $\sim 3\,\mathrm{km\,s^{-1}}$. Show that it would *not* be possible to detect the presence of the Earth from a distant solar system using the Doppler method. State one method that might allow the Earth to be detected. (The Earth's mass is $1/333\,000$ that of the Sun. $1\,\mathrm{AU} = 1.49 \times 10^{11}\,\mathrm{m}$.)

3.    A planet has been discovered orbiting the sun-like star HD 209458. From the period and velocity measured by the Doppler method, its mass was calculated to be 0.69 times that of Jupiter. The plane of the planet's orbit is edge-on and it has been observed to transit across the face of the star. The light from the star was seen to drop by 1.7% when the planet's disc lay fully within that of the star which has a radius of $8 \times 10^5$ km. Assuming that the brightness of the star is constant across the observed face, estimate the diameter of the planet and hence it density in comparison with that of Jupiter which has a radius of $7.1 \times 10^4$ km.

# Chapter 5

# Observing the Universe

This chapter will show how, over the centuries, telescopes observing across the whole of the electromagnetic spectrum have been developed so enabling us to study the universe with ever increasing ability.

## 5.1 Thinking about optics in terms of waves rather than rays

Textbooks and optical design programs describe the working of an optical system, be it a simple lens or a complex telescope, in terms of ray tracing: where the path of a light ray is refracted at the surface of a lens or reflected at the surface of a mirror. Ray tracing implies that, if all the rays traced through an optical system pass through one point, then the system will give a good focus. A simple example is the comparison of a parabolic mirror with a spherical mirror as shown in Figure 5.1. However, in the author's view this is not quite the whole truth and a focus, as given by ray tracing, is a 'necessary but not sufficient' requirement for producing a good image.

### 5.1.1 The parabolic mirror

The fundamental requirement is to do with coherence. Imagine a wavefront arriving from a source, say a star (this is an astronomy book!) that is effectively at infinity. In this case the wavefront will have no curvature and is called a plane wavefront. What does this mean? Simply that light everywhere across the wavefront left the star at the same time and is said to be coherent. You may have seen a wave represented by a rotating vector which rotates once as the wave advances one wavelength through space. The angle of the vector, which conventionally rotates in an anticlockwise sense, defines the phase of the wave. So, if we split up our plane wavefront into little bits – we could call them wavelets – we could represent each wavelet as a rotating vector with each wavelet having the same phase. If, through an optical system, we can bring all these wavelets together to

*Introduction to Astronomy and Cosmology*   Ian Morison
© 2008 John Wiley & Sons, Ltd

Spherical Mirror          Parabolic Mirror

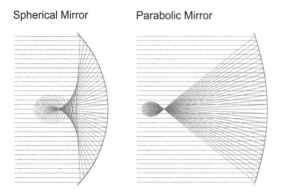

**Figure 5.1 Comparison of a spherical mirror and a parabolic mirror showing that, in the case of the parabolic mirror, all paraxial rays pass through one point (the focus).**

one point *each having travelled for the same time from a plane wavefront* (and hence the source) then the vectors of the wavelets will all have the same phase and will add coherently (in a vector addition) to give a true focus. At other locations within the optical system the wavelets will add with varying phases so that the resultant will, in general, tend to zero. Figure 5.2 using a parabolic mirror will hopefully make this clear. At the focus all the vectors are in phase so the vector addition gives a large resultant but away from the focus the resultant is effectively zero. (The mirror shown has a very short focal length and is representative of the surface of a radio telescope, but the principle is just the same for optical mirrors.)

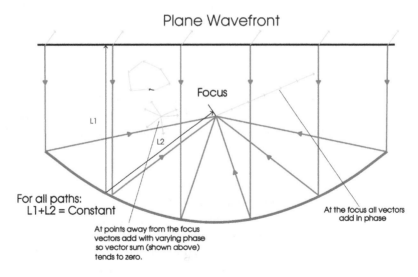

**Figure 5.2 The vector addition of wavelets at, and away from, the focus of a parabola.**

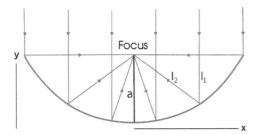

**Figure 5.3** The geometry of a parabola.

So why does a parabolic mirror give a good focus? The answer is simply because a parabolic surface has the property that ($l_1$) the distance from any point on a plane perpendicular to the optical axis to the point on the surface below plus ($l_2$) the distance from this point on the surface to the focus is constant.

A parabola is defined by:

$$y = (1/4a)\, x^2$$

where $a$ is the distance of the focus from the origin (Figure 5.3).

The path length from the plane wavefront across the aperture first to the surface and then to the focus is given by $l_1 + l_2$.

$$l_1 = a - y$$
$$l_2 = [x^2 + (a-y)^2]^{1/2}$$

We can substitute for $x^2$ ($= 4ay$) and expand the square term giving:

$$l_2 = (4ay + a^2 - 2ay + y^2)^{1/2}$$
$$= (a^2 + 2ay + y^2)^{1/2}$$
$$= [(a + y)^2]^{1/2}$$
$$= (a + y)$$

So

$$l_1 + l_2 = (a - y) + (a + y)$$
$$= 2a$$

So the path length for all wavelets to the focus is a constant – they will add coherently.

In order to give its theoretical resolution – when the optics of a mirror or radio telescope surface are termed 'diffraction limited' – the mirror must conform to the

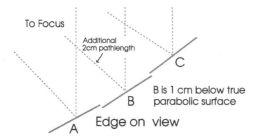

To Focus

Additional
2cm pathlength

C

B is 1 cm below true
parabolic surface
B

Edge on view

A

**Figure 5.4** The effect of an error in the surface of a parabolic surface.

precise parabolic shape to a very high degree of accuracy. Suppose, as shown in Figure 5.4, a small part of a parabolic surface lies below its correct position by a distance of 1/8th of a wavelength (this would correspond to a radio wavelength of 8 cm). The extra path length that the waves falling on this segment will travel to the focal point will be approximately twice this distance (2 cm or 1/4 of a wavelength). The phase of the waves reflected by this segment will thus be $360/4 = 90°$ out of phase with the waves that have been reflected by the surface which lies on the correct parabolic profile. The vector summation at the focus will thus give a smaller resultant – all the waves reaching the focus are not quite coherent and, as a result, the image quality is reduced. A professional mirror or radio telescope surface will probably have errors less than 1/20th of the shortest wavelength that is to be observed.

One problem with a parabolic mirror is that, though it gives a perfect image at the focal point which lays on the axis of the parabola, away from the axis the image quality breaks down due to an optical aberration (called coma) – stars appear to look like little comets. The area of the sky that can be clearly seen is limited by this effect which is why other more complex designs have been developed as will be described later.

### 5.1.2    *Imaging with a thin lens*

Let us now consider what are called *thin* lenses using as an example a biconvex lens made from crown glass. It is *not* possible to prove that there is constant path length from a plane wavefront perpendicular to the optical axis to the focus as it is not true but, for thin lenses, it is very nearly true.

If the focus (where the wavelets add coherently) is the point where all path lengths from a plane wavefront passing through the lens have taken the same time to traverse, then it must be true for the two paths (ray 1) through the centre of the lens and (ray 2) a path that just passes through the tip of the lens where we will assume the additional delay by passing through the glass is zero (Figure 5.5).

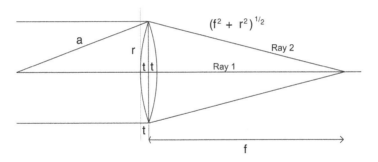

**Figure 5.5** The geometry of imaging with a biconvex lens.

You can immediately see that a lens works by delaying the light passing through its centre by just the same amount as the delay caused by the extra path length in space that the edge ray has to traverse. The delay is due to the fact that the effective path length through glass is increased over that in a vacuum by a factor which is given by the refractive index of the glass, $n$. This varies with different glass types and also, of extreme importance, with the wavelength of the light passing through it – a phenomenon called dispersion. The refractive index of crown glass at the wavelength of green light is typically $1.5$. (Air essentially acts like a vacuum – it scarcely delays light having a refractive index of just $1.0008$ so we will assume that the thin lens is in a vacuum.)

Simple geometry equating the time it takes to traverse the two extreme paths and calculating the thickness of the lens in relation to its diameter and radius of curvature derives a formula giving the focal length as a function of the radius of curvature of the lens and the refractive index of the glass. The resulting equation is exactly that derived by ray optics and is the simplified form of the lensmaker's equation when the radii of curvature of the two lens surfaces are the same.

From Figure 5.5:

$$(f^2 + r^2)^{1/2} + t = (f - t) + 2nt$$
$$= f + t\,(2n - 1)$$
$$(f^2 + r^2)^{1/2} = f + t\,(2n - 2)$$
$$(f^2 + r^2)^{1/2} = f + 2t(n - 1)$$

Square both sides:

$$f^2 + r^2 = f^2 + 4ft(n - 1) + [2t(n - 1)]^2$$

We can ignore the final term. For a thin lens $t$ is very much less than $f$ so the squared term in $t$ is far smaller that the two other parts of the right-hand side of the equation.

In addition we can cancel out the $f^2$ term giving:

$$r^2 = 4ft(n-1)$$
$$f = r^2/4t(n-1)$$

If the radius of curvature of the lens surface is $a$, then:

$$a^2 = (a-t)^2 + r^2$$

Expanding the right-hand side gives:

$$a^2 = a^2 - 2at + t^2 + r^2$$

We can again ignore the $t^2$ terms and cancelling out the $a^2$ terms gives:

$$2at = r^2$$

That is:

$$a = r^2/2t$$

Substituting for $r^2/2t$ in the expression for $f$ gives:

$$f = a/2(n-1)$$

**This is the lensmaker's equation for a thin biconvex lens whose surfaces have the same radii of curvature.**

Table 5.1 gives the refractive indices of both a crown and a flint glass for three colours of light. We will first consider a biconvex lens made of crown glass, and choose a radius of curvature, 1000 mm, which gives a focal length typical of an astronomical refracting telescope.

$$f_{blue} = 954\,mm$$
$$f_{green\text{-}yellow} = 967\,mm$$
$$f_{red} = 970\,mm$$

**Table 5.1** The refractive indices of crown and flint glass at three wavelengths.

|  | Blue<br>486.1 nm | Green-yellow<br>589.3 nm | Red<br>656.3 nm |
|---|---|---|---|
| Crown | 1.524 | 1.517 | 1.515 |
| Flint | 1.639 | 1.627 | 1.622 |

It is immediately apparent that the single lens gives red and blue focuses that are widely separate from the green-yellow focus. Using such a lens in a telescope and trying to get the best image – probably where the green light comes to a focus – one would see the sharp green image superimposed on out of focus blue and red images. Blue and red together make purple, so the in focus image is surrounded by a purple glow. This effect is called chromatic aberration. A small diameter lens with a relatively long focal length can give a passable image and we should not forget that Galileo was able to show that the planets orbit the Sun by making some superb observations of Venus with just such a telescope.

### 5.1.3    *The achromatic doublet*

Combining a biconvex crown glass lens with a plano-concave flint glass lens, gives what is called an achromatic doublet or achromat – implying that the problem of chromatic aberration is largely eliminated. Such a doublet lens was first patented by John Dolland in 1758 but it is believed that the first achromatic lenses were made by Chester Moore Hall in about 1733.

A similar calculation to that above can show how the combination of lenses used by Dolland can markedly reduce the chromatic aberration. We will assume that the plano-concave flint lens has zero thickness in the middle as any additional path length through the flint glass due to its minimum thickness is common to both the central and extreme rays and so will cancel out. (Figure 5.6)

Equating the path lengths of the central and extreme rays gives:

$$f + 2n_1 t = t + n_2 t + (f^2 + r^2)^{1/2}$$
$$f + t(2n_1 - n_2 - 1) = (f^2 + r^2)^{1/2}$$

Squaring both sides gives:

$$f^2 + 2ft(2n_1 - n - 1) + [t(2n_1 - n_2 - 1)]^2 = f^2 + r^2$$

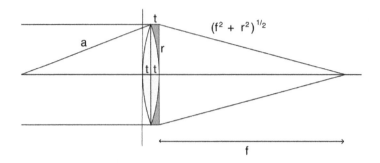

**Figure 5.6** An achromatic doublet.

Cancelling $f^2$, ignoring the term in $t^2$ and simplifying gives:

$$f = r^2/2t(2n_1 - n_2 - 1)$$

Substituting for $r^2$, as above, gives:

$$f = a/(2n_1 - n_2 - 1)$$

The calculation shows that the focal length now depends on $(2n_c - n_f - 1)$. We can choose a value of $a$, 393.6 mm, to give the same focal length for green-yellow light as for the biconvex lens so the results can be directly compared. From Table 5.1 this gives for blue, green-yellow and red, respectively:

$$2n_c - n_f - 1$$

| | | |
|---|---|---|
| Blue | 0.409 | 962.3 mm |
| Green-yellow | 0.407 | 967.0 mm |
| Red | 0.408 | 964.6 mm |

One can see that the spread in focal length is reduced so helping to eliminate the chromatic aberration. This was for a generic pair of glasses. If the pair of glasses is chosen with care, particularly if one has a very high refractive index or, even better, a lens made of calcium fluorite, the correction can be extremely good and the resulting doublet lenses effectively colour free. Such lenses are called apochromats and give superb images. Apochromats may also constructed using a combination of three lenses of differing glass types.

Telescopes using an objective lens, either an achromat or an apochromat, are widely used by amateur astronomers in sizes up to ~150 mm diameter (above 150 mm the cost rises markedly). A pair of binoculars is made up of two short focal length refractors linked together. These employ prisms to fold their light paths and so reduce their overall length as well as to give an erect image.

Many astronomical telescopes now use a combination of lenses and mirrors and these are called catadioptric telescopes. All three types will be described below after a consideration of some general properties of telescopes and why they are employed.

A telescope is used to do two things:

(1)    To collect more light so that fainter objects can be seen better than with the unaided eye.
(2)    To enable us to see greater detail in a distant object.

We will consider each in turn. However, it is important to first understand a little about the human eye which is, after all, a wonderful optical instrument.

## 5.2    The human eye

Our eyes are all that we need if we wish to observe the outlines of the constellations, the majestic sweep of the Milky Way or the fleeting trail of a meteor arcing across the sky. However, to use them to view the heavens at their best, a little time is required for them to become dark-adapted. First, and most obviously, in dark conditions the pupils will dilate so allowing more light to enter the eye (Figure 5.7). This happens over a period of 20 s or so and will thus almost immediately enable one to see fainter stars. The typical daylight size of the pupil is about 2.5–3 mm across but this extends to 5–7 mm under dark conditions. Sadly, as one gets older, the maximum size is likely to be nearer 5 mm rather than 7 mm (a factor of 2 in area) so younger people will definitely be able to see better in the dark!

There is a second dark adaptation mechanism that takes about 20 min to come into effect. Without high light levels reaching the retina, vitamin A is converted first into retinene and then into rhodoposin (visual purple) that significantly improves the sensitivity of the rods and cones which are the light sensitive receptors in the retina of the eye. Strong white light can quickly

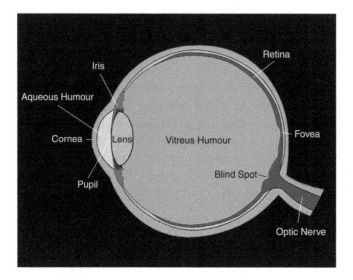

**Figure 5.7** The human eye.

reverse this change, but red light is far less damaging which is why astronomers use red light to view their star charts. In daylight we tend to use a region of the retina called the fovea which is densely populated with cones that are colour sensitive but require high light levels to work well. This is why objects that can have beautiful colours when imaged photographically tend to appear as shades of grey when viewed through a telescope. In contrast, the rods are not colour sensitive but are more sensitive to light. Away from the central foveal region, the cones are fewer in number and the rods are more closely packed. We can thus see fainter objects by using 'averted vision' – directing our gaze to one side of the object of interest.

Even after dark adaptation, what we will actually be able to see on any given night depends critically on the amount of dust and water vapour in the atmosphere. These absorb and scatter the starlight so making it difficult to see fainter objects. The dust and vapour will also scatter back towards our eyes any light from the ground – called light pollution – making the sky appear brighter and the problem even worse. We use the term 'transparency' to define how clear the sky is. At a dark site with a very transparent sky we would typically be able to see a 6–6.5 magnitude star looking towards the zenith. Light pollution combined with an atmosphere laden with dust and water vapour will reduce this value significantly. To give the best chance of viewing the heavens one should try to be as far away from built up areas as possible – even a few miles will help significantly.

## The use of a telescope or pair of binoculars to see fainter objects

Collecting the light with a larger aperture than our eye will allow us to observe fainter objects as the larger aperture of, say, a telescope will collect more light that our dark adapted eye. This increase over the eye alone will depend on the relative areas. A telescope with an objective whose aperture is, say, 150 mm will collect $(150/7)^2$ times more light than a pupil of aperture 7 mm. This is 460 times more light, so one would expect to be able to see a star 460 times fainter that with the eye alone. This can be converted into a magnitude difference:

$$\Delta m = 2.5 \times \log_{10}(460)$$
$$= 6.65$$

Thus we would expect that a fully efficient 150 mm telescope would enable one to see objects 6.6 magnitudes fainter that with the unaided eye. Assuming that our eyes alone could see a star of 6.5 magnitude, then with the 150 mm telescope we might expect to see a star of $6.5 + 6.6 = 13.1$ magnitudes. This is called the limiting magnitude for that telescope.

This calculation relates to telescopes which are 100% efficient, that is, all the light that enters the telescope reaches the eye. This is effectively true for refracting telescopes which may have an efficiency of 98%, but reflecting telescopes will not perform quite as well. The mirrors of a reflecting telescope may only reflect 86% of the incident light so, as two mirrors are required, the efficiency drops to 74%. Modern multi-layer coatings have brought up the reflectivity to 97%, so giving an overall efficiency of nearly 94% – a major improvement. One needs to increase the diameter of the objective by 1.58 to gain an extra magnitude (as from 10 to 16 in.), but the number of visible stars goes up very rapidly with increasing magnitude so even half a magnitude difference will show a lot more stars.

It is not widely known that increasing the magnification of a telescope, discussed later, will enable fainter stars to be seen if there is light in the sky from the Moon or light pollution. Increased magnification reduces the apparent brightness of the overall sky light, but does not reduce the brightness of the stars, so enabling them to stand out more easily. Very good seeing (less turbulence in the atmosphere) and the use of excellent optics will produce smaller stellar images. As their image is spread over a smaller region of the retina the photons are more likely to trigger a response and a star will be seen where under less good seeing it would not. Observing from higher altitudes, there is obviously less atmosphere above and fainter stars will be seen. However, if you go too high, less oxygen reaches the brain and the eyes become less efficient (around 6000–8000 ft is probably the optimum unless you have access to an oxygen bottle). At these heights the unaided eye can perceive stars of 8th magnitude!

## Using a telescope to see more detail in an image

There is a fundamental limit to the detail in the image produced by a telescope which is caused by the effects of diffraction. Assuming that the telescope has a circular aperture, and in the absence of an atmosphere, then the image formed by a point source is a central disc surrounded by a number of concentric rings rapidly decreasing in brightness. In a perfect telescope the central disc contains 84% of the light with the remainder in the surrounding rings – most in the first ring which has about twice the diameter of the disc. This pattern, called the Airy disc, shown in Figure 5.8, is named after the Astronomer Royal, George Airy, who gained some notoriety for not pursuing the discovery of the planet Neptune.

The angular size of this pattern, $\Delta\theta$, is a function of both the wavelength of light, $\lambda$, and the diameter of the telescope objective, $D$:

$$\Delta\theta = 1.22\,\lambda/D$$

The angular size of the Airy disc is related to what is termed the resolution of a telescope. For a given wavelength the resolution increases with telescope aperture and is inversely proportional to $D$.

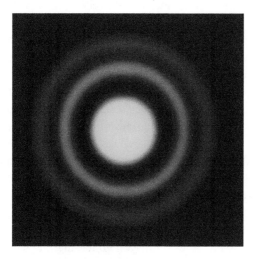

**Figure 5.8** The Airy disc.

The calculation of this formula requires calculus but an approximate derivation can be gained by using a little quantum mechanics. It is based on the Heisenberg Uncertainty Principle which states, in one of its forms, that the more accurately we know the position of an object, the less accurately we can know its momentum. The principle states that the product of the uncertainty in $x$, $\Delta_x$, times the uncertainty in $p$, $\Delta_p$, must be less that Planck's constant, $h$. When a photon passes through an aperture (like a lens) its position is determined to a certain extent so we cannot be so sure about its momentum along the axis or axes in which it is constrained. This is at right angles to the direction of the incoming photon and the result is to give the photon some uncertainty in its future direction. A beam of individual photons passing through the aperture thus 'spreads out' giving rise to the Airy disc. You can immediately see that the smaller the aperture through which the photons pass, the greater the knowledge about their position and hence the less well determined is their momentum and the Airy disc gets larger – agreeing with the formula given above.

Consider a photon passing through a slot aperture along the $x$ axis which has a width $D$ in the $y$ axis as shown in Figure 5.9. It has come from infinity along the $x$ axis and has a momentum $p$. Quantum mechanics states that the momentum, $p_x$, of the photon is given by:

$$p_x = h/\lambda$$

where $h$ is Planck's constant and $\lambda$ is the wavelength.

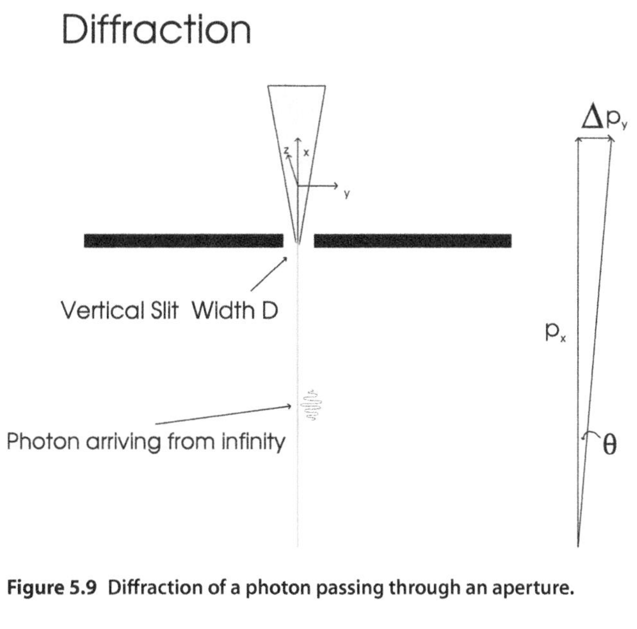

## Diffraction

**Figure 5.9** Diffraction of a photon passing through an aperture.

The fact that it has passed through the slit will give the photon an uncertainty in its momentum in the $y$ direction of $\Delta p_y$, where as $\Delta x = D$:

$$\Delta p_y D = h$$

From the momentum equation, $h = p_x \lambda$,

So,    $$\Delta p_y D = p_x \lambda,$$

Giving    $$\Delta p_y / p_x = \lambda / D.$$

The uncertainty in $\Delta p_y$ gives an uncertainty in the direction in which the photon continues of $\Delta \theta = \Delta p_y / p_x$ so, substituting from above gives:

$$\Delta \theta = \lambda / D.$$

This is the angular uncertainty when the photon's position has only been constrained along one axis. A lens of aperture $D$ will constrain the photons along two axes, and one would expect the uncertainty to become greater. Hence the factor 1.22 derived by the formal calculation.

### 5.4.1    *An interesting worked example of the effects of diffraction*

In Section 3.5.4 it was pointed out that the lunar laser reflectors were made up of 100 small reflectors rather than one large one. Let us see why. The laser pulse sent from a telescope on Earth is reflected by 100, 3.8 cm square, corner-cube reflectors forming a square array whose sides are 46 cm in length.

Consider first a single 46 cm square reflector. The light pulse reflected by the reflector could be thought of as a beam of light coming from *beyond* the Moon's surface that has passed through a square aperture of this size on its way to Earth. Assuming that a green laser is used, the beam will thus be spread over an angle given by:

$$\Delta \theta = \sim 1.22 \, \lambda / D$$
$$= \sim 1.22 \times 5.5 \times 10^{-7} / 0.46$$
$$= \sim 1.5 \times 10^{-6} \, \text{rad}$$

The beam will thus be spread over this angle and could thus be seen over an area of the Earth whose diameter, $d$, is given by $R \times \Delta \theta$, where $R$ is the distance of the Moon's surface from a typical point on the surface of the Earth.

$$d = \sim 1.5 \times 10^{-6} \times 378\,000$$
$$= \sim 0.57\,\text{km}$$

The radius of this area will thus be ~0.28 km.

However during the 2.5 s the pulse takes to reach and return from the Moon, the telescope will have moved as the Earth rotates around its axis. If the telescope that sent the pulse was located at a latitude of 40° then the distance moved, $d_t$, would be given by the circumference of the Earth at latitude 40 times the ratio of 2.5 s to 1 day:

$$d_t = 2 \times \pi \times [6780 \times \cos(40)] \times 2.5/24 \times 3600$$
$$d_t = 0.94\,\text{km}$$

So the laser ranging telescope will have moved out of the region where it could receive the returned echo!

By using smaller individual reflectors, in this case 3.8 cm, the reflected beam will have an angular size 46/3.8 times larger and, on the Earth, will cover an area of radius 3.4 km which is comfortably greater than the telescope's change of position.

## 5.4.2 The effect of diffraction on the resolution of a telescope

If one considers a 150 mm telescope observing in green light of $5.5 \times 10^{-7}$ m wavelength, one gets a size for the Airy disc of $4.4 \times 10^{-6}$ rad which is 0.9 arcsec:

$$\Delta\theta = 1.22\,\lambda/D$$
$$= 1.22 \times 5.5 \times 10^{-7}/0.15\,\text{rad}$$
$$= 4.4 \times 10^{-6}\,\text{rad}$$
$$= 4.4 \times 10^{-6} \times 57.3 \times 3600\,\text{arcsec}$$
$$= 0.9\,\text{arcsec}$$

Larger aperture telescopes will theoretically give higher resolutions but, in practice the resolution is usually limited by what is called the 'seeing' – a function of turbulence in the atmosphere. The atmosphere contains cells of gas with slightly differing refractive indices which are carried high above the telescope by the wind and act rather like the glass used for screens which blur what is seen beyond. A star is effectively a point source and should theoretically have an image size given by the Airy disc of the telescope aperture. In practice a stellar image size as seen from the UK will probably be or order 2–3 arcsec across and is highly unsteady. This is one reason why professional telescopes are located on high mountains on islands such as La Palma in the Atlantic Ocean and Hawaii in the Pacific Ocean. At such locations, there is far less atmosphere above the telescope and the air tends to be less turbulent as it has been flowing over the sea. Under the best conditions the seeing might limit the resolution to half an arcsecond, so larger aperture telescopes *will* see more detail but not significantly more than a telescope whose aperture is ~400 cm across. The best location for an optical telescope is in space, as in the case of the Hubble Space Telescope, where its full resolution of 1/20th of an arcsecond at visible wavelengths may be realised.

The turbulence of the atmosphere and hence seeing varies from night to night. In bad seeing the image of a star will appear bloated and the Moon can appear to be boiling! On such nights the image quality will be totally determined by the atmosphere. But, rarely, the atmosphere can be still and then the aperture, type of telescope and the quality of the optics will determine what you can see. The seeing tends to have more effect on large aperture telescopes, so when the seeing is not good a smaller telescope may actually give better views of the planets.

## 5.5    The magnification of a telescope

The objective of a telescope, be it lens or mirror, produces an image in its focal plane. Here a charge-coupled device (CCD) array might be placed to produce an image of the object or the image might be viewed by using an eyepiece. The eyepiece works like a magnifying glass enabling one to view the image in detail. Eyepieces have focal lengths ranging from ~2.5 up to ~55 mm. The magnification of a telescope is simply obtained by dividing the focal length of the objective by the focal length of the eyepiece, so a 12 mm focal length eyepiece coupled with a 1200 mm focal length objective gives a magnification of $\times 100$. It is also possible to place a concave lens in the light path close to the eyepiece. This is referred to as a Barlow lens, and has the effect of diverging the light cone and giving the effect of having an objective of greater focal length, so providing an effective increase in magnification for a given primary focal length. The effective focal length is usually doubled (called $\times 2$ Barlows), but can be obtained with effective magnification ratios of 2.5 or even 4.

Eyepieces generally give excellent performance on axis, but image quality drops off the further off axis one observes. The designer of an eyepiece will determine the diameter of the area that can be observed with the eyepiece without significant loss of detail and use a 'field stop' to limit the observed field. This then determines the area of sky that can be observed when used with a given objective.

Suppose that there is an object that lies away from the centre of the field of view by an angle of $\theta/2$ that appears right at the edge of the image encompassed by the field

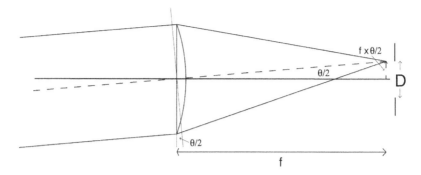

**Figure 5.10** The field of view of a telescope.

stop of the eyepiece. As Figure 5.10 shows, the image of this object will lie a distance of $(f \times \theta/2)$, where $\theta$ is in radians, from the centre of the image. If the diameter of the field stop of the eyepiece is $D$, then $D = 2 \times f \times \theta/2 = f \times \theta$. So, the diameter of the field of view, $\theta$, of a telescope whose focal length is $f$ mm coupled with an eyepiece whose field stop has a diameter $D$ mm, is given by $D/f$ rad. As an example, consider a 1200 mm objective when used with an eyepiece whose field stop is 12 mm across. Then $\theta$ is $1200/12 = 1/100$ rad or $57.3/100° = 0.573°$. This combination would thus nicely encompass an image of the Moon which is ~0.5° across.

The diameter of the field stop is approximately proportional to the focal length when eyepieces of similar design are used, so using a shorter focal length eyepiece gives greater magnification at the expense of observing a smaller field of view. The Plossl eyepiece, made up of two achromatic doublets, is very widely used as a 'standard' eyepiece. Eyepieces using five or more elements can allow wider field stops for a given focal length and so allow a larger field of view to be observed. These are called 'wide field' eyepieces and can be very expensive.

One might well think that as the magnification is increased more detail will be seen. This is not necessarily so and it is very rare that magnifications greater than ×200 will be useful. Both the eye itself and the atmosphere play a part. The detail in an image observed by the eye is limited by two things: the diffraction effects due to the limited aperture of the eye and the density of the rods and cones in the retina. The achievable resolution is limited by the aperture of the pupil, say 3 mm, which gives an Airy disc whose size, $\Delta\theta$, is given by:

$$
\begin{aligned}
\Delta\theta &= 1.22\,\lambda/D \\
&= 1.22 \times 5.5 \times 10^{-7}/0.003\,\text{rad} \\
&= 2.24 \times 10^{-4}\,\text{rad} \\
&= 2.24 \times 10^{-4} \times 57.3 \times 60\,\text{arcmin} \\
&= 0.77\,\text{arcmin}
\end{aligned}
$$

Thus the eye has a theoretical resolution of about 1 arcmin (1/60th of a degree). This image falls upon the retina where it is sampled by the rods and cones. If there were an insufficient number of these, the resolution would be compromised, but evolution has ensured that this is not a limiting factor! If we have a telescope with a magnification of ×60 it means that we could see detail about 1 arcsec across. It is rare that the atmosphere will allow us to see detail finer than this so one might say that ×60 is all that you need. In fact, spreading out the image on the retina helps so a magnification of ×120 – ×200 will usually let us see more detail.

<table>
<tr><td>**5.6**</td></tr>
</table>

## Image contrast

In any telescope, some light will fall in areas of the image where it should not – turning black into dark grey. This light has been scattered by the surfaces of lenses and mirrors in the telescope and stray light can be reflected from the sides of the telescope tube. Mirrors are worse than lenses in this respect and it is harder to efficiently baffle (to intercept stray light) a reflecting telescope than a refractor. So refractors tend to give the highest contrast images. Many manufacturers can now provide mirrors with high reflectivity coatings. These will allow slightly fainter objects to be seen but, more importantly, by reducing the light scattered at mirror surfaces, they will have a marked effect on image contrast.

<table>
<tr><td>**5.7**</td></tr>
</table>

## The classic Newtonian telescope

The Newtonian telescope was invented by Sir Isaac Newton who did not believe that one could overcome the problem of chromatic aberration that was suffered by simple refracting telescopes of the time. Sadly, it is not thought that he made any astronomical observations with it. In the Newtonian, a primary mirror reflects the light to a focus that would lay in the centre of the tube so, to avoid obstructing the light path with the head, a secondary mirror, often called the flat, reflects the light sideways to form an image just outside the tube where the focuser and eyepiece are placed. As the secondary mirror has to intercept the converging light cone at an angle of 45° it must have an elliptical outline with its major axis 1.414 (the square root of 2) times longer than its minor axis.

In Figure 5.11, let us assume that the focal plane is 10 mm outside the telescope tube which has an external diameter of 230 mm and which houses a primary mirror of focal length 1600 mm and diameter 200 mm. The centre of the secondary mirror is thus at a distance of (10 + 115) mm from the focal plane, that is 125 mm. One can then calculate the *minimum* length of the minor axis

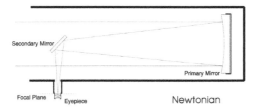

**Figure 5.11** The Newtonian telescope.

using similar triangles. If $D$ is the diameter of the primary mirror, $d$ is the minor axis of the secondary mirror, $F$ is the focal length of the primary mirror and $k$ is the distance of the image plane from the telescope axis, then:

$$D/F = d/k$$

With the given values:

$$d = D \times k/F$$
$$= 200 \times 125/1600$$
$$= 15.6 \, \text{mm}.$$

The corresponding major axis is $15.6 \times 1.414 = 22 \, \text{mm}$.

Why does this give the minimum size of the secondary mirror? Imagine putting the pupil of your eye at the focal point of the telescope. You would be able to see the whole of the primary mirror surface. However, if you move your eye to one side you will no longer be able to see the whole of the surface as the secondary mirror will not be large enough. The effect of this is to reduce the effective area of the primary mirror for image points away from the axis, causing what is called vignetting of the image away from the axis. The actual size of the secondary mirror is thus increased to minimize this effect, the amount dependent upon the main use of the telescope.

As described above, the 'image' of a star is a disc surrounded by a number of circles, the so-called Airy disc. With an unobstructed aperture (as with a refractor) little light (16%) falls in the rings but, as the secondary mirror of a reflecting telescope gets larger, more light falls into the rings and the effect is to reduce the effective resolution of the telescope (though the diameter of the central part of the disc actually gets smaller). The secondary mirror causes a second problem – it has to be supported. Usually this is done with a four vane 'spider'. This is what causes the 'cross', formed by diffraction spikes, seen around bright stars in many images you may have seen. Thus light is moved away from where it should be and so the image is slightly degraded.

## 5.8    The Cassegrain telescope

The majority of professional telescopes are a variant of a type of telescope called a Cassegrain. In these telescopes the secondary mirror is a hyperboloid which reflects the light down through a central hole through the primary mirror to the focal plane. This is a far better place to locate heavy equipment such as a spectrometer. It has been pointed out that the image quality of a parabolic primary falls off rather rapidly away from the optical axis due to the optical aberration called coma. If, instead, the primary is also a hyperboloid in what is called a Ritchey–Chrétien telescope, coma is eliminated making it well suited for wide field observations. (This telescope was invented by George Willis Ritchey and Henri Chrétien in the early 1910s.) The vast majority of all professional telescopes, including the Hubble Space Telescope, are of this design.

## 5.9    Catadioptric telescopes

This class of telescopes uses a combination of a mirror and a lens to produce the image. In general, spherical mirrors (which are cheap to make) are used to produce the image together with the lens – normally called a corrector lens – to correct for the spherical aberration that would be caused by the spherical mirror.

### 5.9.1    *The Schmidt camera*

The Schmidt camera was invented in 1930 by Bernhard Schmidt. He wanted to design a new type of instrument which would have a very large field of view yet be free of aberrations such as coma and would have a short focal ratio so allowing fainter stars to be observed for a given exposure time. To eliminate chromatic aberration the new design would use a mirror as the primary, but to give a large field of view and eliminate coma a spherical mirror would be needed. However, spherical mirrors suffer from spherical aberration. He realized that he could correct for spherical aberration if a corrector plate was placed at the radius of curvature of the spherical mirror. This has a varying thickness across its aperture to compensate for the fact that a spherical, rather than a parabolic, mirror is used.

Schmidt cameras have become one of the most useful tools of modern astronomy, ideally suited to photographing large star fields in the Milky Way, showing 10 000 stars on one negative (Figure 5.12)! Highly valuable sky surveys have been made using such cameras, one example being the 48 in. Samuel Oschin Schmidt Telescope at Mount Palomar which produced the Palomar Sky Survey, completed in 1958. The film plates were 14 in. square and covered an area of sky 6° across! The survey initially covered the whole of the northern sky down to a

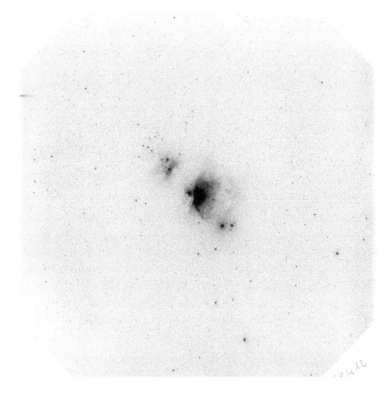

**Figure 5.12** Schmidt plate showing the Orion Nebula region. Image: Asaigo Science Archive Database.

declination of minus $-27°$. Plates were made with both blue and red sensitive emulsions and were sensitive to stars of $+22$ magnitudes (about 1 million times fainter than the limit of human vision).

On a personal note, an asteroid numbered 15 727, now named after the author, was discovered in 1990 by the 2 m diameter Alfred-Jensch-Telescope at Tautenburg Observatory in Germany, the largest Schmidt camera in the world.

**5.9.2**    *The Schmidt–Cassegrain telescope*

The design of the Schmidt camera has led to a widely used variant of the Cassegrain telescope. In larger sizes, parabolic mirrors are expensive to make and the length of their telescope tubes make them rather unwieldy. Both of these issues are addressed in the Schmidt–Cassegrain design which uses a spherical mirror in a compact optical tube assembly (Figure 5.13). The spherical aberration that would ruin the image quality of the spherical mirror is corrected by a 'Schmidt' corrector plate at the front of the telescope. This also holds the secondary mirror which reflects the light back

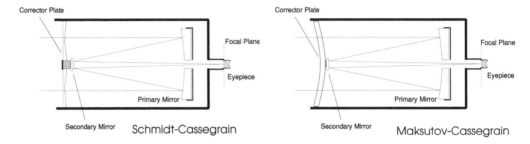

**Figure 5.13** Catadioptric telescopes.

through the primary as in the 'Cassegrain' design. There are thus no diffraction spikes (as caused by the spider in a Newtonian telescope) and, as the whole tube is enclosed, the mirror is kept dust free. The majority of Schmidt–Cassegrain telescopes have a focal ratio of *f*10 resulting in relatively long focal lengths. (A Schmidt–Cassegrain with a primary mirror of 200 mm will have a focal length of 2000 mm.) A consequence of the design is that quite a large 'central obstruction' is required; partly as a relatively large secondary mirror is required by this design, but also to help prevent light entering the telescope aperture passing directly into the eyepiece. As in a Newtonian telescope, the central obstruction also has an effect on the image quality; light being transferred from the central peak of the Airy disc into the surrounding rings – though the diameter of the central disc is actually reduced.

### 5.9.3    *The Maksutov–Cassegrain telescope*

The Maksutov is a catadioptric telescope design, developed by the Russian optical specialist Dmitri Maksutov in 1941, which employs a full diameter meniscus lens to correct the problems of off-axis aberrations. They are usually of a Cassegrain form, with the secondary mirror either silvered on the inner face of the corrector lens or supported by it as shown in Figure 5.13. The result is a very compact telescope that gives a relatively wide field of view and excellent quality images. As large aperture corrector plates are difficult to manufacture, their professional application is limited.

## 5.10    Active and adaptive optics

Given a perfect telescope used in space, resolution is directly proportional to the inverse of the telescope diameter. A plane wavefront from a distant star would form an image, with an angular resolution only limited by the diffraction of light and the telescope is said to be diffraction limited. However, in practice, turbulence in the atmosphere distorts the wavefront, creating phase errors across the mirror. Even at the best sites, ground-based telescopes observing at visible wavelengths cannot achieve an angular resolution better than telescopes of about 20 cm diameter. Wavefront errors are also caused by inaccuracies in the mirror's surface and effects caused by gravitational and thermal changes in the mirror and its support structure.

The wavefront errors are thus of two types:

(1) Slowly changing errors due to gravitational and thermal effects on the mirror. These are corrected by active optics systems.
(2) Rapidly changing errors due to the turbulence in the atmosphere. These are corrected by adaptive optics systems.

## 5.10.1 Active optics

This is the term used to describe the methods used to correct for slow changes in the mirror and its telescope structure. In a typical system the mirror – which is relatively thin – is supported by a large number of actuators (perhaps 150 in number) which can be moved to apply forces to the rear surface of the mirror and so adjust the surface profile. Periodically the image of a star is analysed, taking about 30 s, and a computer calculates the errors in the surface that would give rise to the observed image. The computer then calculates the correction which each actuator has to apply to achieve optimal image quality.

## 5.10.2 Adaptive optics

This is the term used to describe the correction of phase errors caused by the atmosphere. Across a large mirror, rapidly changing phase errors equivalent to a few micrometres path length result. If there is a reasonably bright star in the field of view (which should give an image which is that of an Airy disc whose angular size is determined by the aperture of the telescope) its actual, distorted, image can be analysed in a computer and corrections applied to correct for the waveform errors and so produce a diffraction limited image. It would be quite impossible to correct the primary mirror at the required rate of every few milliseconds to the required precision of ~1/50th of a micrometre so, instead, a small (8–20 cm) deformable mirror is used in the light path whose surface profile is controlled by hundreds of piezoelectric actuators which move to compensate for the atmospheric effects. In the near infrared (as the wavelengths are longer) it is easier to fully correct the wavefront and so give a diffraction limited performance, but such systems can still provide an improvement of perhaps ten times at visible wavelengths.

Often a suitable reference star will not be found in the field of view that includes the target object, so artificial reference stars are now being created by firing a laser which excites sodium atoms in the upper atmosphere at an altitude of ~90 km. Such an artificial reference star can be created as close to the astronomical target as desired, but a lookout has to be kept for high flying aircraft!

## 5.11  Some significant optical telescopes

### 5.11.1  *Gemini North and South telescopes*

The Gemini Observatory comprises two 8.1-m telescopes; Gemini North (Figure 5.14) is located on Mauna Kea, Hawaii at a height of 4214 m whilst Gemini South is at a height of 2737 m on Cerro Pachón, Chile. Together, the twin telescopes can give full sky coverage with both sites giving a high percentage of clear weather and excellent atmospheric conditions. They have been designed to operate especially well at infrared wavelengths and to this end, their mirrors are coated with silver which reflects significantly more infrared radiation than the aluminium used to coat most other telescope mirrors. As atmospheric water vapour absorbs infrared radiation both telescopes are located on high mountain tops where the air has a very low water vapour content.

**Figure 5.14** The Gemini North telescope. Note the open sides of the dome to allow the telescope to remain in thermal equilibrium with the outside air. Image: Neelon Crawford, Polar Fine Arts, courtesy of Gemini Observatory and National Science Foundation.

### 5.11.2     *The Keck telescopes*

The twin Keck telescopes, located at a height of 4200 m at the top of Mauna Kea, Hawaii, are the world's largest optical and infrared telescopes (Figure 5.15). They have primary mirrors of 10 m diameter each composed of 36 hexagonal segments whose positions are adjusted (using active optics) to act as a single mirror. Telescopes can take time to thermally stabilize to the night-time temperatures when the domes are opened, so to minimize these effects, the interior of the Keck domes are chilled close to freezing point during the day.

When observing, twice a second when observing, the active optics system controls the positions of each mirror segment to a precision of 4 nm to compensate for thermal and gravitational deformations. The Keck telescopes also use an adaptive optics system using 15 cm diameter deformable mirrors that changes their shape up to 670 times per second to cancel out atmospheric distortion, improving the image quality by a fact of ten.

The Keck telescopes have made a notable contribution to the detection of extra-solar planets by the 'radial velocity' method.

### 5.11.3     *The South Africa Large Telescope (SALT)*

With a spherical 10 m mirror made up of 91 hexagonal segments, SALT is the largest telescope in the southern hemisphere and its innovative design is based on the Hobby-Ebberly Telescope at the McDonald Observatory in Texas (Figure 5.16). The telescope is tilted at a fixed angle of 37° from the zenith and moves only in azimuth, rotating on air bearings to point to the region of sky that is to be observed.

**Figure 5.15**  The twin Keck telescopes. Image: Courtesy W. M. Keck Observatory.

**Figure 5.16** The South Africa Large Telescope. Image: SALT Consortium.

During the observation the telescope structure remains stationary whilst an optical corrector assembly (to correct for the spherical aberration of the mirror) and instrument payload move across the top of the telescope tube to track the object being observed. The design allowed SALT to be built at less than a fifth of the cost of a conventional 10-m telescope.

### 5.11.4    *The Very Large Telescope (VLT)*

The VLT is operated by the European Southern Observatory (ESO) and consists of four 8.2-m telescopes which can either work independently or in a combined mode when it is equivalent to a single 16-m telescope – making it the largest optical telescope in the world (Figure 5.17). The light from the four auxiliary 1.8-m telescopes may also be combined with that from the 8.2-m telescopes to give high angular resolution imaging. It can observe over a wavelength range from the near ultraviolet up to 25 μm in the infrared.

The VLT is located at the Paranal Observatory on Cerro Paranal in the Atacama Desert, northern Chile, at a height of 2600 m (one of the best observing sites in the world). The four main telescopes have been given the names of objects in the sky in the Mapuche language: Antu (the Sun), Kueyen (the Moon), Melipal

**Figure 5.17** The four 8.2-m telescopes of the VLT, with the four 1.8-m auxiliary telescopes in the foreground. Image: Courtesy of the European Southern Observatory.

(the Southern Cross) and Yepun (the star Sirius). One notable achievement was the first visual image of a planet, albeit around a brown dwarf rather than a normal star, which was observed in the infrared when using an adaptive optics system.

## 5.11.5    *The Hubble Space Telescope (HST)*

The HST (Figure 5.18) was launched in April 1990 to observe the universe over a wavelength range that extends from the ultraviolet, through the visible to the near infrared; that is, from 0.12 μm in the ultraviolet to 2.4 μm in the near infrared (1 μm is $10^{-6}$ m).

The HST's primary mirror is 2.4 m across so that in green light its angular resolution is given by:

$$
\begin{aligned}
\Delta\theta &= 1.22\,\lambda/D \\
&= 1.22 \times 5.1 \times 10^{-7}/2.4 \,\text{rad} \\
&= 2.5 \times 10^{-7} \,\text{rad} \\
&= 3.4 \times 10^{-3} \times 57.3 \times 3600 \,\text{arcsec} \\
&= 0.053 \,\text{arcsec} \\
&= {\sim}1/20\text{th of an arcsecond}
\end{aligned}
$$

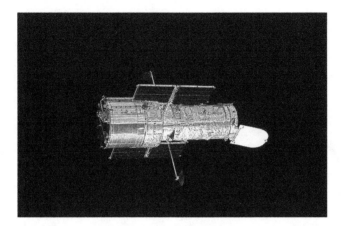

**Figure 5.18** The Hubble Space Telescope. Image: Space Telescope Science Institute, NASA.

However, this resolution would have only been met if the mirror was ground to such a precision that it would be diffraction limited and to achieve this, the mirror was one of the smoothest ever made.

When placed in its orbit some 600 km above the Earth, the astronomers commissioning the HST were mortified to find that it could not be focused. The full optical system had not been tested on the ground and a problem in the test rig that controlled the shape of the primary mirror meant that the mirror suffered significant spherical aberration. The resultant image quality was very poor. The centre of the mirror was just 2 μm too shallow and, as a result, rays from the edge of the mirror came to a focus 4 cm behind that from the centre. An ideal mirror will put 84% of the light into the central disc, but the HST mirror put only ~15% of the light in the central disc with most of the light spread over a region 1.5 arcsec across, comparable with ground based telescopes!

However, rather than being a badly made mirror of the correct shape, it was a perfectly made mirror of the wrong shape! This meant that it was possible to correct the spherical aberration by introducing a correcting lens or mirror into the optical path, and this was achieved in the first servicing mission. It is now a virtually perfect telescope and its contribution to our understanding of the universe has been immense. One of its main legacies to astronomy will be the images of the distant universe in what are called the Hubble Deep Field and the Hubble Ultra Deep Field showing galaxies close to the time of their formation just a billion years after the origin of the universe.

## 5.11.6    *The future of optical astronomy*

Plans are being made for a number of very large optical telescopes to be constructed in the twenty-first century which will have unprecedented sensitivity. One such concept being considered is OWL, the design for a 100-m optical

**Figure 5.19** Artist's impression of the 100-m optical telescope concept. Image: European Southern Observatory.

telescope that would have the ability to observe details down to 1 milliarcsec. It would observe in the open air and be covered by a dome during daylight hours as shown in Figure 5.19. Due to the very high cost envisaged it is perhaps more likely that an optical telescope in the size range 30–60 m would be built in the coming decades. There is also the long term prospect of building and operating an optical telescope on the Moon – where there is no atmosphere to degrade image quality and no light pollution!

## 5.12    Radio telescopes

These are used for the wavelength range that typically ranges from 0.75 m (408 MHz) down to 1 cm (22 GHz). This is the range of wavelengths that reach the ground through the 'radio window' limited at the bottom end by the ionosphere and at the upper end by absorption by water vapour in the atmosphere. A great advantage of the fact that radio waves are far longer than optical wavelengths is that they are not absorbed by dust. Optical astronomers can only see ~10% of the way towards the centre of our Milky Way Galaxy due the thick dust lanes that lie along its plane. Radio astronomers really can observe parts of the universe inaccessible to optical astronomers!

The majority of modern telescopes are either prime focus paraboloids, where the receiver is located at the prime focus above the surface, or of a cassegrain design, which uses a secondary mirror to reflect the radio waves to a point below the centre of the primary reflecting surface to a point where the radio receivers are located. They tend to have very short focal ratios as in Figure 5.1.

### 5.12.1    *The feed and low noise amplifier system*

At the focus, a feed collects the radio waves that have been focused by the primary reflector (and possibly a secondary reflector) and passes the signal to a low noise amplifier made using transistors with inherently low noise levels. However, at room temperature, the electrons within the amplifier produce a significant amount of thermal noise which can be reduced by cooling the amplifiers to temperatures near to absolute zero. The amplifiers are mounted within a cryostat that is cooled by a gaseous helium refrigeration system. To minimize the heat reaching the amplifier from its surroundings, the interior of the cryostat is evacuated to eliminate heat transfer by convection, a radiation shield of highly polished metal reduces transfer of heat by radiation and the wires to supply current to the amplifiers are carried by fine wires wrapped around the coldest part of the refrigeration system to extract heat within them. The receivers then reach temperatures of ~12 K and the thermal noise is reduced to minimal levels.

However, the noise produced within the low noise amplifier is only part of the total noise that the radio telescope will receive even when not directly pointing at a radio source. The total noise is determined by the 'system noise temperature', which is the temperature of a cavity in which a noiseless receiver is placed which would give the same signal strength as the real system whose noise contributions come from a number of causes (Figure 5.20).

In summary, these are:

(1)    The noise produced by the cooled amplifier; this might be of order 8 K.

(2)    Noise left over from the Big Bang. This is called the Cosmic Microwave Background, and has an equivalent temperature of ~3 K; it will be discussed in detail in Chapter 9.

(3)    Noise from our own Milky Way Galaxy, caused by electrons travelling close to the speed of light (called relativistic electrons) spiralling around the magnetic field of the Galaxy. It is called synchrotron radiation as it was first observed being emitted from particle accelerators called synchrotrons.

(4)    Radiation from molecules in the atmosphere, water vapour being a particular problem. The contribution will be less when the humidity is low, and high dry locations will always give a lower contribution. Higher frequencies are particularly affected, which is why Jodrell Bank's radio telescopes observing the Cosmic Microwave Background at 30 GHz (~0.7 cm wavelength) are located on the flanks of Mount Teide on Tenerife

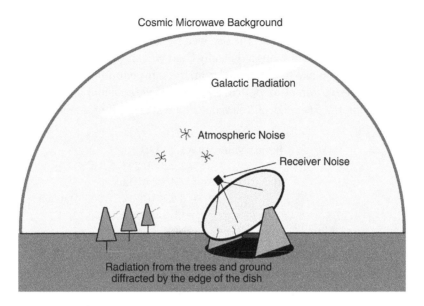

**Figure 5.20** The contributions to the total system noise which is typically 25 K.

in the Canary Islands. Antarctica is a superb location from which to observe at very short wavelengths as the water vapour is frozen out of the atmosphere!

(5) The ground is at ~290 K and so radiates as a black body at this temperature. Of course the radio telescope does not look at the ground, but some radiation can be diffracted around the edge of the dish, and the struts that support the secondary reflector will also scatter radiation into the feed system of the telescope.

The higher the frequencies transferred down coaxial cables from the focus the greater their loss, so the frequencies are translated downwards to lower frequencies (called down-conversion to intermediate frequencies) before transfer along cables to the laboratories where the data acquisition systems are located. In the more modern systems, the radio frequency data is sampled after initial amplification and then transferred along fibreoptic cables carried by modulated infrared laser light.

### 5.12.2    *Radio receivers*

The simplest form of radio receiver detects the radio signal and measures the received signal strength. Such systems initially plotted out the data on chart recorders but now, of course, computers digitize and record the data. These can be used to carry out surveys of the radio sky. As the radio telescope scans across a strong radio source, the received signal strength will rise to a maximum and then fall, plotting out a typically Gaussian shape on the chart. This is the beam width that can be calculated from the same equation that was used to calculate the resolution of an optical telescope. If we consider a typical scenario, that of using the 76-m Lovell Telescope at a wavelength of 21 cm then:

$$
\begin{aligned}
\text{Beam width} = \Delta\theta &= 1.22\,\lambda/D \\
&= 1.22 \times 0.21/76\,\text{rad} \\
&= 3.4 \times 10^{-3}\,\text{rad} \\
&= 3.4 \times 10^{-3} \times 57.3 \times 60\,\text{arcmin} \\
&= 11.6\,\text{arcmin}
\end{aligned}
$$

You can immediately see that the effective resolution of even the third largest fully steerable telescope in the world is ~12 times less than the human eye at visible wavelengths! In fact this is an underestimate of the actual beam width as the feed of the antenna will not collect waves reflected from towards the edge of the dish as efficiently as from its centre. This means that the effective diameter of the dish is smaller than its actual size, and this can be compensated for by using a larger constant in the equation for the beam width. A value of 1.6 rather than 1.22 is appropriate, giving a beam width for the Lovell Telescope at 21 cm wavelength of ~15 arcmin.

### 5.12.3    *Telescope designs*

The Lovell Telescope is a prime focus design, where the receiver is placed directly at the focus of the primary reflector (Figure 5.21). However, most modern telescopes use a Cassegrain configuration as in Figure 5.22. One advantage is that in the region of the focal point, which is behind the primary reflector, it is possible to locate a 'carousel' supporting a number of receivers each of which can be brought to the focal point by simply and quickly rotating the carousel. This gives what is termed 'frequency flexibility'. It should be pointed out that a Cassegrain design is not well suited for low frequencies as the size of the secondary reflector has to become prohibitively large. In some telescopes the secondary mirror can be replaced by a primary focus box, and in others a dielectric 'lens' is used above the feed at the Cassegrain focus to improve the efficiency. At very low frequencies, a primary focus system is mandatory; which is why NASA, whose space tracking radio telescopes are of Cassegrain design, asked for the use of the (prime focus) Lovell Telescope in attempts to receive signals directly from two space probes, lost at Mars, which should have been transmitting at the frequencies just above 400 MHz.

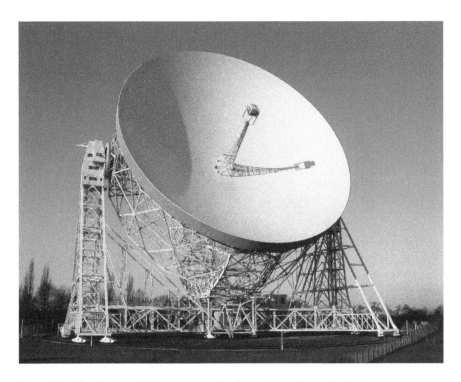

**Figure 5.21** The 76-m Lovell Telescope – a prime focus design. Image: Ian Morison.

The shape of large parabolic antennas such as the 100-m Effelsberg Telescope in Germany and the 105-m Green Bank Telescope in West Virginia, USA will distort as their elevation changes due to changes in the gravitational forces acting on the structure (Figure 5.23). Both telescopes employ techniques to mitigate these effects. The Effelsberg Telescope uses a novel method which allows the shape to distort, but always to a surface with an accurate parabolic shape. The position of the focus will vary in elevation so the feed is moved to follow the focal point. In the Green Bank Telescope each of the 2004 panels that make up the surface area is mounted on actuators located at each of their corners. A laser system, mounted close to the focus, is used to measure the precise position of each panel and to adjust them as necessary to compensate for the gravitational deformations. (This is equivalent to the technique of active optics that is used for optical telescopes.) The swan neck holds the secondary mirror and receiver systems away from the path of radio waves falling on the primary reflector. This is called an offset-feed system. It prevents the supports that hold the secondary reflector and receivers scattering noise into the system.

**Figure 5.22**  The 32-m Cambridge Telescope – a Cassegrain design. The cone houses a carousel supporting four receivers. Image: Ian Morison.

### 5.12.4    *Large fixed dishes*

Currently the world's largest radio telescope is the 305-m Arecibo Telescope in Puerto Rico (Figure 5.24). It is built within a natural depression in the ground and has a spherical surface. High above the dish is a complex system of secondary and tertiary mirrors in the receiver complex to correct for the inherent spherical aberration. The secondary mirror is itself 25 m in diameter! As a radio source moves across the sky it is 'tracked' by moving the secondary mirror and receiver complex along an arm suspended above the surface.

A telescope called FAST is being built in China which will have an overall diameter of 500 m. Each of the 4600 panels that will make up the surface will be mounted on actuators so that, as a radio source moves above the telescope, part of the surface, approximately 300 m across, is formed into a parabolic surface to reflect incident radio waves up to a focus above the surface. The focus box is moved above the surface to follow the focal point by controlling the length of the support cables. This provides a coarse adjustment, with a laser controlled system to provide fine control of the feed position within the focus box. This innovative telescope should be completed by 2013.

**Figure 5.23**  The 105-m Robert C. Byrd Green Bank Telescope. Image: National Radio Astronomy Observatory.

**Figure 5.24**  The 305-m Arecibo Telescope. Image courtesy of the NAIC – Arecibo Observatory, a facility of the NSF.

### 5.12.5    *Telescope arrays*

The resolution provided at radio wavelengths by even the largest single dishes is far below that of even a small optical telescope. By linking a number of antennas into an array it is possible to synthesize a telescope whose effective size is given by the greatest separation of antennas in the array. Consider a line of antennas 5 km long whose central antenna was located at the North Pole. Looking down from above, this line of telescopes would be seen to rotate and, in 12 h, would fill out a circular area of diameter 5 km. If the data received by all of the antennas during the 12 h period is processed in an appropriate way an image of the sky above can be produced whose resolution is equal to that of a 5-km telescope. As this depends on the rotation of the Earth the technique is termed 'Earth Rotation Aperture Synthesis'. It is obviously not practical to locate an array at the North Pole, but providing one is not too near the equator the technique is viable.

At each instant of observation, each pair of telescopes is measuring some information about the structure of the radio source depending on their separation and orientation relative to the radio source. Due to the rotation of the Earth, the information about the radio source changes throughout the observing period and this enables an image of the source to be created. If there are $N$ antennas in the array there are $N(N-1)/2$ combinations. The signals from each antenna are brought to a central 'correlator'. Here delays have to be introduced into the signal path of each telescope so that the signals are combined coherently.

Three significant arrays are:

- The Very Large Array (VLA) in the plains of New Mexico (Figure 5.25). It comprises three arms of nine 25-m antennas in a 'Y' formation. The array can be configured in a number of ways with a greatest overall separation of 36 km. Waveguides are currently used to carry the radio signals to the central 'correlator' where the signals from the individual telescopes are combined in pairs. It can operate at wavelengths of 400 cm down to 0.7 cm.
- The MERLIN array across the UK has an overall size of 217 km and normally incorporates five 25-m antennas and one 32-m antenna. For observations where high sensitivity is required the 76-m Lovell Telescope can be incorporated into the array. It can operate at wavelengths of 74 cm down to 1.3 cm.
- The Giant Metre Radio Telescope (GMRT) is located at a site near Pune in India. The GMRT consists of thirty 45-m diameter dishes separated by up to 25 km. Fourteen of the 30 dishes are located in a compact central array in a region of about 1 km² with the remaining 16 dishes spread out along the three arms of a 'Y'-shaped configuration. It can operate over a wavelength range of 6 m down to 21 cm.

**Figure 5.25** The Very Large Array in New Mexico, USA. Image: Wikipedia Commons.

The resolution of these arrays depends on the overall diameter and operating wavelength. One significant attribute of the MERLIN array is that its resolution is comparable with that of the HST. At the wavelength of 5 cm the resolution is given by:

$$\text{Resolution} = \Delta\theta = 1.22\, \lambda/D$$
$$= 1.22 \times 0.05/217\,000\,\text{rad}$$
$$= 2.8 \times 10^{-7}\,\text{rad}$$
$$= 2.8 \times 10^{-7} \times 57.3 \times 3600\,\text{arcsec}$$
$$= 0.06\,\text{arcsec}$$

This is 1/15th of an arcsecond, comparable with that of the HST resolution at visible wavelengths and so allows images in the optical and radio parts of the electromagnetic spectrum to be directly compared.

## 5.12.6     *Very Long Baseline Interferometry (VLBI)*

The highest resolutions available in any part of the electromagnetic spectrum are provided by arrays of radio telescopes that cross continents in what are called VLBI arrays. The core of the European VLBI Network (EVN) incorporates antennas stretching from the UK across to Poland and from Finland to Sicily. It is also possible to include two antennas in China, one in South Africa and the Arecibo Telescope in the Caribbean. As a radio source is observed simultaneously at each observatory, the data is digitized and 'time stamped' using the observatory's atomic clock. The data are then stored on arrays of hard disks. At the end of the observing session the hard disks are shipped to Holland where the data are combined in a correlator at JIVE – the Joint Institute for VLBI in Europe. It is now also possible to transfer data directly to JIVE using the internet and this allows immediate analysis of the data which is a great asset when planning observations of transient events. Due to the great size of the array, resolutions of less than 1 milliarcsec are possible allowing, for example, the expanding shell of a supernova in a galaxy 12 million light years to be imaged and its expansion rate measured (as shown in Figure 5.26). The white spots in the radio image, lower right, are supernovae remnants, one of which has been imaged by the EVN over a 12-year interval.

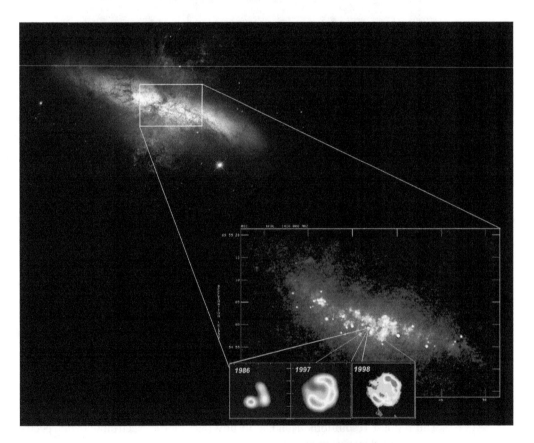

**Figure 5.26** The starburst galaxy M82. Upper left: Hubble Space Telescope image of M82 in visible light. Lower right: VLA and MERLIN radio image of the central part of M82 with, below, EVN images of an expanding supernova remnant. Image: Hubble Space Telescope Institute, NASA and the University of Manchester.

Two further VLBI arrays exist; the US Very Long Baseline Array (VLBA) that stretches from Hawaii to St Croix in the Caribbean, and the Asia Pacific Telescope (APT) which incorporates antennas in Australia, Japan, China and South Africa to provide high resolution studies of radio sources in the southern hemisphere.

### 5.12.7    *The future of radio astronomy*

The international radio astronomical community is now building and planning the radio telescopes that will dominate the observations of the twenty-first century – the Atacama Large Millimetre Array (ALMA) and the Square Kilometre Array (SKA). ALMA is being constructed in the Atacama Desert region of Chile at a height of 5000 m (Figure 5.27). Here there is little water vapour in the atmosphere above so enabling observations from 10 mm down to as little as 0.3 mm wavelengths. The 50 or more individual antennas can be spaced up to several kilometres apart giving unprecedented resolution at these short wavelengths. ALMA will be able to observe the clouds of gas and dust from which all the stars and galaxies in the universe form and help us to more fully understand the process of star formation.

**Figure 5.27**  Artist's impression of the Atacama Large Millimetre Array. Image: European Southern Observatory.

The SKA will be built in either South Africa or Australia and will have an effective collecting area of $1 \, km^2$ – hence its name. It would thus have the same collecting area as 130, 100-m radio telescopes! The collecting area of the main array will, however, be made up arrays of small antennas, expected to be 15 m in size. This requires the use of more electronic systems but overall, the required collecting area can be achieved more cheaply than by the use of larger individual antennas.

The SKA will consist of an inner core of antennas making up a large part of the array's total collecting area, together with outer stations arranged in a log-spiral pattern extending out to distances of 3000 km. It will thus combine high sensitivity allied to high resolution to give an instrument whose performance will greatly exceed radio telescope arrays currently in use. The SKA will be a highly flexible instrument designed to address many of the most fundamental questions in astronomy. It will be the only instrument capable of observing the universe in the period, called the 'dark ages', when the hydrogen and helium formed in the Big Bang had yet to form the first stars. Close to the central core will be a 'phased array' of receiving elements capable of observing much of the sky at one time – ideal for detecting transient events – and also able to carry out several observing tasks simultaneously. The artist's impression of the SKA in Figure 5.28 shows the central core at the upper right, some of the outer stations in the foreground and middle left, the phased array. Phase 1 of the project, with over 10% of the total collecting area, is hoped to be completed by 2015 with final completion of the array in 2020.

**Figure 5.28** Artist's impression of the central region of the SKA. Image: Xilos Studios, SKA Project Development Office.

## 5.13     Observing in other wavebands

### 5.13.1     *Infrared*

Many large optical telescopes are also used to observe in the infrared region of the spectrum, but there are some dedicated telescopes, such as the United Kingdom Infrared Telescope (UKIRT) located at a height of 4194 m on Mauna Kea, Hawaii (Figure 5.29). The UKIRT has a 3.8-m mirror and is equipped with a new mid infrared eschelle spectrometer (Michelle); it has been used to observe young stars which were previously hidden in cocoons of the dust and gas from which they formed. Observations at 'near infrared' wavelengths cannot penetrate the surrounding dusty material, but Michelle's view at longer 'mid infrared' wavelengths reveals the young stars within.

UKIRT has also significantly advanced our understanding of brown dwarfs, mysterious objects sometimes referred to as 'failed stars'. They are more massive than gas giant planets like Jupiter, but are not quite massive enough to shine like normal stars.

### 5.13.2     *Submillimetre wavelengths*

Submillimetre wavelengths lie between infrared light and radio waves on the wavelength scale. This largely comes from cold material in the universe, such as the clouds of gas and dust found between the stars that form the 'interstellar medium'. It is out of this dust and gas that new stars are born, and into which stars disperse material as they explode at the end of their lives.

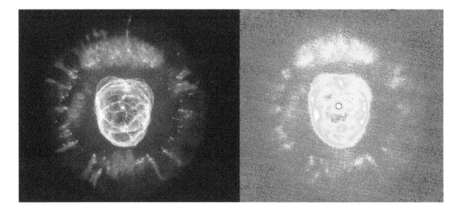

**Figure 5.29** The Eskimo planetary nebula imaged in visible light by the Hubble Space Telescope (left) and in the infrared by the United Kingdom Infrared Telescope (right). Image: Hubble Space Telescope Institute, NASA and Chris Davis and Tim Carroll (Joint Astronomy Centre).

**Figure 5.30** The James Clark Maxwell Telescope. Image: Science and Technology Facilities Council.

The James Clark Maxwell Telescope (JCMT) is the largest submillimetre telescope in the world having a 15-m diameter dish and is located on Mauna Kea, Hawaii (Figure 5.30). (You may have noticed that it is pretty crowded up there!)

As is the case for infrared, water vapour in the Earth's atmosphere absorbs submillimetre waves – which is why the high, dry site of Mauna Kea is one of the best places on the planet for astronomy.

The JCMT is equipped with a detector array called SCUBA, the world's most powerful submillimetre wave camera which takes pictures showing the faint heat radiation of interstellar dust grains. These fine particles, like soot or sand, are at temperatures below $-240°C$. To be able to detect such faint radiation, SCUBA itself is kept within a jacket of liquid helium at a temperature of less than 1/10th of a degree above absolute zero! The JCMT also has 'heterodyne receivers' which detect light from gas molecules in space; their submillimetre radiation tell us about the temperature, density, and motion of the gas.

In the nearby massive star formation region, the Orion Nebula, SCUBA's maps have revealed bright knots where stars are being born and has shown a complex region of shells, filaments, and clouds at the centre of our Galaxy hidden in visible light by intervening dust. SCUBA can also observe galaxies, enshrouded in dust, more than 10 billion light years away and give information about star birth in the early universe.

### 5.13.3     *The Spitzer space telescope*

At wavelengths between 3 µm and 180 µm, most of this infrared radiation is blocked by the Earth's atmosphere and cannot be observed from the ground so NASA has launched the Spitzer Space Telescope which has a 0.85-m telescope and three cryogenically cooled science instruments. Spitzer is the largest infrared telescope ever launched into space and, like SCUBA in the JCMT, the telescope must be cooled to near absolute zero so that it can observe infrared signals from space without interference from the telescope's own heat. To protect the telescope from the heat of the Sun and infrared radiation from the Earth, Spitzer carries a solar shield and its orbit, trailing that of the Earth, places Spitzer far enough away from the Earth to allow the telescope to cool rapidly without having to carry large amounts of cryogen (coolant).

### 5.13.4     *Ultraviolet, X-ray and gamma-ray observatories*

To observe at these very short wavelengths, the telescopes have to be in space and NASA has launched two space observatories, the Compton Gamma-ray Observatory and the Chandra X-ray Observatory, to observe at the shortest wavelengths, whilst the HST can observe at ultraviolet wavelengths,

#### Ultraviolet

There are currently no dedicated ultraviolet observatories in orbit, but the HST carries out significant observing at ultraviolet wavelengths, but it has a fairly small field of view. From 1978 until September 1996, the International Ultraviolet Explorer (IUE) was operating to observe ultraviolet radiation. There have, however, been some ultraviolet space missions where a set of three wide field telescopes, called Astro, were carried into orbit in the bay of the space shuttle.

#### Extreme ultraviolet observatories

Astronomers have been somewhat reluctant to build space telescopes to observe at extreme short ultraviolet wavelengths since theories suggests that the interstellar medium (the tiny traces of gases and dust between the stars) would absorb much of this radiation in this part of the spectrum. However, when the Extreme Ultraviolet Explorer (EUVE) was launched, observations showed that the region around the Sun is relatively sparse of gas and dust – it appears to be located within a bubble in the local interstellar medium – and this enabled the EUVE instruments to see much further than expected.

### The ultraviolet/X-ray boundary

An array of low energy X-ray imaging sensors (ALEXIS) was launched in 1995. Though its name implies that it is an X-ray observatory, the range of wavelengths observed by ALEXIS is at the very lowest end of the X-ray spectrum and often considered to be extreme ultraviolet.

### X-rays – The Chandra X-ray observatory

NASA's X-ray observatory is named in honour of Subrahmanyan Chandrasekhar who, at the age of 19, determined the mass limit for white dwarf stars. It was launched and deployed by the Space Shuttle Columbia in 1999 for a nominal 5-year mission.

Mirrors cannot be used to focus X-ray radiation as X-ray photons would be absorbed by normal mirror surfaces. Instead X-ray telescopes use nested cylindrical paraboloid and hyperboloid surfaces coated with iridium or gold, deployed so that the X-rays are incident at very low grazing angles. (An analogy is the fact that a water surface reflects light well when it falls on it at a very shallow angle.) Chandra uses four pairs of nested iridium mirrors to give a resolution of 0.5 arcsec.

Chandra observations have revolutionized the field of X-ray astronomy: the image of the supernova remnant Cassiopeia A, gave a first glimpse of the compact object at its centre, probably a neutron star or black hole (Figure 5.31). X-ray

**Figure 5.31** The Chandra X-ray Observatory with inset showing the X-ray image of Cassiopeia A. Image: Chandra, NASA.

emission has been seen from the region surrounding the super-massive black hole, Sagittarius A*, at the centre of the Milky Way and Chandra has made a measurement of Hubble's constant giving a value of $76.9 \, \text{km} \, \text{s}^{-1} \, \text{Mpc}^{-1}$. Recently, Chandra's observation of colliding superclusters has found strong evidence for the existence of dark matter.

### Gamma rays – The Compton Gamma-ray observatory

The Compton Gamma-ray Observatory, named after Dr Arthur Holly Compton, a Nobel Prize winner for work involved with gamma-ray physics, was launched in April 1991 and was deployed in low earth orbit at 450 km in order to avoid the Van Allen radiation belt.

It carried a complement of four instruments to cover gamma-ray energies from 20 keV to 30 GeV. A key use of the telescope was the detection of transient events, called gamma-ray bursts (GRBs). It detected, on average, 1 GRB per day with a total of 2700 detections. A GRB observed in 1999, one of the brightest bursts recorded, was the first GRB with an optical afterglow that allowed astronomers to measure a redshift of 1.6 for the object giving rise to the gamma rays which corresponds to a distance of 4.5 Gpc. It appears that GRBs result when a massive star explodes in what is called a hypernova or perhaps when two neutron stars coalesce.

Following the failure of one of its control gyroscopes, the observatory was deliberately brought out of orbit and its debris fell into the Pacific Ocean in June 2000.

## 5.14  Observing the universe without using electromagnetic radiation

Cosmic rays and gravitational waves can provide a way to learn about the Universe in other ways than by observing electromagnetic radiation.

### 5.14.1  *Cosmic rays*

Cosmic rays are energetic particles that enter the Earth's atmosphere from space. Protons make up nearly 90% of all incoming cosmic ray particles with about 9% helium nuclei (alpha particles) and about 1% electrons along with a small component of more massive nuclei. The particles arrive individually, not in a beam, so the term 'ray' is a misnomer. Some are highly energetic with energies of over $10^{20}$ eV, considerably more than we can produce with particle accelerators here on Earth. To put this in context, the Large Hadron Collider which came into operation in

2008 can produce protons with just $7 \times 10^{12}$ eV. The highest energy cosmic rays have an energy equivalent to that of a well served tennis ball!

We believe that most cosmic rays originate from galactic objects such as rotating neutron stars, supernovae, and black holes but the most energetic are thought to come from objects such as the regions surrounding a super-massive black hole at the heart of a distant galaxy. These are called active galactic nuclei and will be discussed in Chapter 8. At the lower end of their energy scale are those originating from energetic processes within the Sun's atmosphere – their numbers increasing markedly after a solar flare.

Cosmic ray particles are divided into two groups: the primary particles, such as oxygen and carbon nuclei produced within stars and ejected in supernova explosions, can interact with the interstellar medium and produce lighter secondary particles, such as the nuclei of lithium, beryllium and boron. The rate at which cosmic rays are detected on Earth has a dependence on the interaction of the solar wind with the Earth's magnetic field. As a result, the rate depends on the solar cycle as the solar wind intensity increases at times of sunspot maximum. Some scientists believe that the extent of the Earth's cloud cover may have a dependence on the cosmic ray flux and so changing solar activity may have an effect on global temperatures.

When cosmic ray particles interact with oxygen and nitrogen in the Earth's atmosphere they produce a cascade of lighter particles, called a cosmic ray air shower which can contain billions of particles. In a significant reaction, neutrons can collide with a $^{14}$N nucleus to give a proton and $^{14}$C nucleus. Arising from this process, the amount of $^{14}$C in the Earth's atmosphere has been kept constant for at least 100 000 years. (It increased in the latter part of the last century due to the testing of nuclear bombs.) Carbon dioxide containing a $^{14}$C atom thus makes up a small part of the carbon dioxide that will be absorbed, for example, by a growing tree. As a result, wood initially contains a well defined ratio of $^{14}$C and $^{12}$C in its makeup. However, $^{14}$C is radioactive, having a half-life of 5730 years, with $^{14}$C decaying to form $^{14}$N with an electron and an antineutrino. The ratio of $^{14}$C to $^{12}$C will thus decrease over time. This enables the age of the wood to be found, a process known as radiocarbon dating.

In their interactions in the atmosphere, cosmic rays can produce pions and kaons which quickly decay into muons. These do not interact strongly with the atmosphere and many reach the surface as a result of the effect of time dilation as described in Section 1.6.6. At rest relative to us, muons decay into an electron and two neutrinos with a half-life of just 1.6 μs. They are produced at a height of about 10 km in the atmosphere and, travelling at ~0.98% of the speed of light, take 34 μs to reach the ground. If time dilation did not occur, only about 1 in 3 000 000 would reach the ground. However, due to its relativistic speed, the muon's clock, as observed by us, is running slow by a factor of about 5, so its effective half-life is 7.8 μs. As a result, 150 000 rather than a single muon out of an initial 3 000 000 will reach the ground.

In chambers 700 m below ground at the Soudan Mine in Minnesota, an experiment has detected 33 000 000 muons over a 10-year period when the Moon was in the field of view. Cosmic rays are shielded by the Moon, and this led to a reduction of detections from the direction of the Moon. The 'Moon's shadow' is slightly offset from its true position as the cosmic rays, being charged particles, are deflected somewhat by the Earth's magnetic field (Figure 5.32).

### 5.14.2    *Gravitational waves*

As was described in Chapter 1, a consequence of Einstein's General Theory of Relativity is that changing mass distributions transfer information across the universe by the emission of gravitational waves. As will be discussed in Chapter 7, one of the strongest sources of gravitational wave emission would be the coalescence of two neutron stars (each typically weighing more than 1 solar

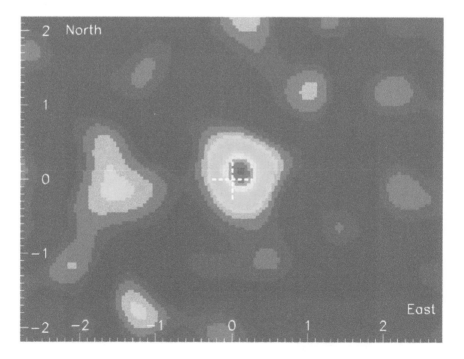

**Figure 5.32** The 'Moon's shadow' as observed by the Soudan 2 muon detector. Image: Argonne National Laboratory, and Minnesota, Tufts and Western Washington Universities, USA, and Oxford University and the Rutherford Appleton Laboratory, UK.

mass) to form a black hole. At the time of writing, though indirect evidence of such emission has been obtained, no gravitational wave has yet been detected by a gravitational wave detector. It is, however, expected that such a direct detection will be made on a timescale of a few years as the current detectors are made more sensitive and so could detect such events across a wide volume of the universe.

The effect of a passing gravitational wave is to produce a tiny distortion of space–time that could give rise to a temporary distortion of a shape, say a sphere, of material by a factor of less than 1 part in $10^{20}$. It is somewhat amazing that such a small effect could be detected!

The simplest, and first to be built, type of gravitational wave detector is called a Weber bar – a large cylinder of metal with electronic devices strapped around it to detect the vibrations induced by a passing gravitational wave. Modern versions are cryogenically cooled and use superconducting quantum interference devices to detect any motion. Unfortunately, Weber bars are not very sensitive but could, for example, detect a nearby supernova explosion. More sensitive detectors use laser interferometers to detect the slight change in the separation of masses placed many hundreds of metres to several kilometres apart. In the USA, at Richland, Washington and Livingston, Louisiana, are two ground-based interferometers forming the Laser Interferometer Gravitational Wave Observatory (LIGO) (Figure 5.33).

**Figure 5.33** The Livingston, Louisiana LIGO gravitational wave detector. Image: LIGO$_{\text{Livingston}}$.

Each comprises two 4 km arms at right angles to each other whose lengths will be changed by a passing gravitational wave by no more than $10^{-17}$ m – only a fraction of the width of a proton! Laser beams pass through vacuum tubes to measure any difference in the separation of test masses at the ends of the arms and enable LIGO's interferometers to detect changes at the level required. Upgrades to LIGO were funded in 2008 to improve its sensitivity by at least a factor of 10 to form Advanced LIGO. For a given event, such as the coalescence of two neutron stars, this will expand the volume in which it could be detected by a factor of 1000. As a result, such events are expected to be detected several times a day.

The passing of a gravitational wave must be detected in the presence of seismic noise, not just from earthquakes, but passing cars and trains and waves reaching the nearby coastline. The test masses are thus suspended by a series of wire frames to filter out seismic noise and the whole assembly is mounted on a support which has accelerometers to measure major disturbances and apply an opposing correction. This works in a very similar way to that employed in modern camera lenses to provide image stabilization.

A space-based interferometer called the Laser Interferometer Space Antenna (LISA) is also being developed (Figure 5.34). Here, three test masses will be placed in spacecraft, 5 million km apart, forming a 60° triangle. Lasers will span the distance between them, detecting the minute changes in their separation as a gravitational wave passes through space. Although LISA will not be affected by seismic noise, it will be affected by other means, such as the Sun's radiation pressure and solar wind,

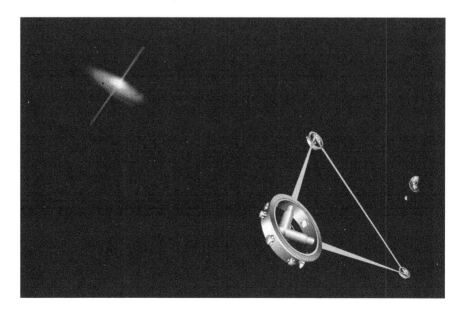

**Figure 5.34** The Laser Interferometer Space Antenna (LISA). Image: NASA.

which would change the relative positions of the test mass. To overcome this, the test masses effectively float free in space protected by their surrounding spacecraft. Additional laser systems control micro thrusters to counteract the forces on the spacecraft and keep its position relative to the test mass accurate to 10 nm – of order 1/100th the wavelength of light!

As we will see in Chapter 9, it is not possible to look back in time further than ~380 000 years after the origin of the universe. Prior to that time, the universe was sufficiently hot for photons to ionize the atoms so filling space with free electrons which scatter light – rather like water droplets in a fog – with the result that the universe was then opaque to electromagnetic radiation. The sensitivity of LISA might actually allow us to probe even further back in time by detecting gravitational waves that were created less than one-trillionth of a second after the Big Bang!

## 5.15   Questions

1.  Stellar magnitudes obey a logarithmic scale with the difference in brightness between a magnitude 1 star and a magnitude 6 star defined to be exactly 100. Given that the eye has a dark adapted pupil size of 5 mm, how large would a telescope aperture have to be in order to observe a star 7 magnitudes fainter than the human eye can detect?

2.  A telescope of 1000 mm focal length is provided with 32 and 11 mm Plossl eyepieces along with a ×2 Barlow lens. What four magnifications are available? The 32 and 11 mm eyepieces have field stop diameters of 27 mm and 9 mm, respectively. What fields of view would be observed with these two eyepieces?

3.  A Newtonian telescope has a mirror diameter of 300 mm and a focal length of 1500 mm. The tube has an external diameter of 330 mm and the focal plane is 40 mm outside the tube.
    (a)  Calculate the focal ratio.
    (b)  Calculate the minor and major axes of the minimum size of secondary mirror that should be used with this telescope.
    (c)  What would be the magnification when used with a 15 mm eyepiece?
    (d)  What would be the field of view when used with an eyepiece having a 44 mm field stop?
    (e)  What is the theoretical resolution, in arcseconds, of the telescope at the wavelength of green light, $5.1 \times 10^{-7}$ m?
    (f)  What tolerance in the manufacture of the mirror surface would you need if the maximum phase error were not to exceed 1/6 $\lambda$ at this wavelength?

4.  The author imaged Jupiter when closest to the Earth last year using his 230 mm aperture, $f10$, Schmidt–Cassegrain telescope with a 170 pixels per millimetre CCD camera having an effective area of $3.9 \times 2.8$ mm. To increase the image scale the effective focal ratio of the telescope was increased by a factor of 2.5 by the use of a Barlow lens.
    (a) What was the effective focal length?
    (b) Calculate the field of view covered by the CCD array.
    (c) Estimate Jupiter's angular size given that it is in a circular orbit of radius 5.7 AU, and has a diameter of 142 800 km. ($1 \text{ AU} = 1.49 \times 10^8$ km.)
    (d) Assuming that Jupiter's equator lay parallel to one axis of the CCD chip, how many pixels lay across Jupiter's image?
    (e) Why do you think that the Barlow lens was used?

5.  Calculate the resolution of the 2.4-m mirror of the Hubble Space Telescope at a wavelength of $0.5 \times 10^{-6}$ m. How large would an array of radio telescopes operating at a wavelength of 0.06 m need to be to equal this resolution?

6.  What would the resolution be of the Square Kilometre Array when operating at 21 cm wavelength using baselines out to 2000 km?

# Chapter 6

# The Properties of Stars

Virtually all of the light that we can see in the heavens has come from the surface of a star and they are, certainly for our existence, the most important objects in the Universe. The heat and light of the Sun sustains life here on Earth, and the life cycles of stars that existed billions of years ago produced the elements of which we are made and that of the planet on which we live. This chapter will discuss the properties of stars in general, how these are measured and where our own star, the Sun, is positioned amongst them.

## 6.1 Stellar luminosity

Stars have a very wide range of intrinsic luminosity – their energy output across the whole electromagnetic spectrum. The luminosity of our Sun, $L_{sun}$, is taken as the reference which, as was calculated in Section 2.2.2, is $3.86 \times 10^{26}$ W.

In Chapter 1 the apparent magnitude scale was described. It was pointed out that this tells us nothing about the relative brightness of stars as an intrinsically very bright star at a great distance might very well have a greater magnitude (i.e., appear less bright) than a nearby intrinsically faint star. Thus, in order to be able to relate the intrinsic brightness and hence luminosity of other stars to that of our Sun we have to eliminate the square law effect of the star's distance. To do this we need to be able to measure the distances of stars.

## 6.2 Stellar distances

To measure stellar distances the method of parallax is used. Suppose that you had to measure the width of a river. Fix on a point directly across the river. Then walk along the river bank until, looking back at it over your shoulder, the point lies at an angle of 45° to the side of the river bank. You will have walked a distance equal to the width of the river. So the method of parallax requires one to observe an object from two positions some distance apart (forming a baseline) and measure the change in angle. The further away the object, the greater the baseline required. In the case of stars, there is no perceptible change in angle from points across the Earth, so a considerably bigger baseline is required. Happily, one is available to us, the distance across the Earth's orbit around the Sun (Figure 6.1). The majority of stars are too far away to show any change in their observed position when observed, say, in the spring and autumn when the Earth is at opposite sides of the Sun. These stars can thus be considered as reference points against which the movement in position of (at present) a relatively few nearby stars can be measured. The angular change in their position, coupled with knowledge of the Earth's orbit, can thus be used to find their distance.

The measured angles are very small, an arcsecond or less, and depend on the exact times at which the measurements were made. To determine the distance, the change in angle of a star's position (which has been measured using a baseline of ~2 AU) is converted into what is termed the **parallax** of the star which is the angular shift in position of the star that *would* be observed with a baseline of exactly 1 AU. As the angles are (very) small, one can then immediately derive its distance from $\theta = A/d$, where $\theta$ is the parallax in radians, $A$ is the astronomical unit and $d$ is the distance. If $A$ is in kilometres, then $d$ would also be in kilometres.

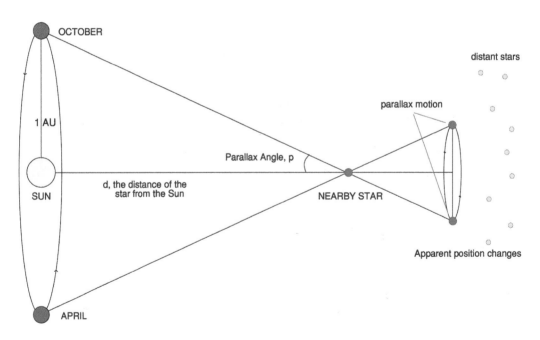

**Figure 6.1** The method of parallax.

As an example, let us consider a star whose parallax is 1/10th of an arcsecond:

$$\theta = 0.1/(3600 \times 57.3)\,\text{rad}$$
$$= 4.85 \times 10^{-7}\,\text{rad}$$

As 1 AU is $1.49598 \times 10^{8}\,\text{km}$, the distance, $d$, is thus given by:

$$d = 1.49598 \times 10^{8}/4.85 \times 10^{-7}\,\text{km}$$
$$= 3.084 \times 10^{14}\,\text{km}$$

One light year is $9.46 \times 10^{12}\,\text{km}$; this is equal to 32.6 light years.

### 6.2.1    *The parsec*

As the angle and distance are inversely related, it is possible to define a unit of distance such that a star located at this distance would have a parallax of 1 arcsec. This unit is called the parsec, and is the unit of distance normally used by professional astronomers. The simple relationship is:

$$d = 1/p$$

where $d$ is in parsecs and $p$ is the parallax of the star in arcseconds. (The parsec is usually abbreviated to pc.)

So a star which subtends an angle of 1/10th of an arcsecond will be at a distance of 10 pc. We have just calculated that this distance is 32.6 light years, so:

1 pc is equal to 3.26 light years

The nearest star has a parallax of 0.772 arcsec which corresponds to a distance $d$ given by:

$$d = 1/0.772\,\text{pc}$$
$$= 1.295\,\text{pc}$$
$$= 1.295 \times 3.26\,\text{light years}$$
$$= 4.22\,\text{light years}$$

It is a star in the α Centauri multiple star system and, not surprisingly, has been given the name Proxima Centauri.

## 6.3    Proper motion

In practice, as is often the case, things are not quite so simple. All stars in the galaxy are moving around the centre of the galaxy and, unless they are moving either directly away or towards us, will be slowly across the sky. This is called a star's proper motion and is usually expressed in units of arcseconds a year. Thus, if one measures a change in the position of a star when observed from either side of the Earth's orbit, one does not know whether this is caused by parallax, proper motion or a combination of both. To separate the two effects, one needs to observe the star again after a full year when the Earth is back to its original position. Any change in position observed over the full year will only be due to proper motion and so this component of the observed motion can be measured, so enabling the part due to parallax to be found. In practice, observations over several years will give the best results.

### 6.3.1    *Hipparcos and GAIA*

We saw in the Chapter 5 that observations from the ground are limited in positional accuracy by turbulence in the atmosphere and, until recently, the number of stars whose parallax was known was limited to the few thousand out to distances of about 40 pc (130 light years). In 1989, unencumbered by the atmosphere, a satellite called Hipparcos (High Precision Parallax Collecting Satellite) was able to make observations for 3 years whose positions were accurate to about a milliarcsecond. It was thus able to accurately measure the distances and proper motions of 118 000 stars out to a distance of ~90 pc (300 light years). The data collected by Hipparcos contributed to the prediction of when comet Shoemaker-Levy 9 would collide with Jupiter and showed that many billions of years ago, the galaxy swallowed a large group of stars.

Even so, Hipparcos was only able to make measurements across ~1% of the size of the galaxy. A 5-year space mission called GAIA is expected to be launched in 2011 which will have an accuracy that is 100 times better than that of Hipparcos (Figure 6.2). GAIA will then be able to measure the positions, distances and proper motions of a billion stars – about 1% of all the stars in the galaxy! Each of these stars and many other types of objects will be observed about 100 times, resulting in a record of their brightness and position over time. Together with the unprecedented accuracy of the astrometric measurements, this will lead to the discovery

**Figure 6.2** Artist's impression of the GAIA spacecraft. Image: ESA.

of planets around other stars as outlined in Chapter 4, asteroids in the inner Solar System and icy bodies in the outer Solar System.

In the meantime, a method called spectroscopic parallax can be used to measure distances across the galaxy. The discussion of this method will follow later in Section 6.8 when stellar spectra have been covered.

## 6.4 The absolute magnitude scale

If all stars were at a fixed distance, then their perceived relative brightness would be a true indication of their relative luminosity. Astronomers used this idea to set up an absolute magnitude scale in which the absolute magnitude of a star is equivalent to the apparent magnitude that it would appear to have if it lay at a distance of 10 pc. The absolute magnitudes of stars closer than 10 pc will thus be *increased* relative to their apparent magnitudes and those further away will be *reduced*.

One problem in measuring the apparent magnitudes is the loss of light through absorption by dust – called extinction. This will underestimate the apparent luminosity (giving a greater apparent magnitude) and the star will be given a smaller absolute magnitude than it should. This problem will be addressed later in the chapter. Ignoring any extinction, if a star had an apparent magnitude of $-20$ at a distance of 100 pc, then at a distance of 10 pc it would appear $10^2$ times brighter (the inverse square law). However, a factor of 100 is exactly 5 magnitudes difference, so its absolute magnitude will be $(-20 + 5)$ that is, 15th magnitude.

### 6.4.1    *The standard formula to derive absolute magnitudes*

Given the distance in parsecs, what is the absolute magnitude, $M$, of a star with apparent magnitude $m$?

The brightness ratio of the star as would be seen at a distance of 10 pc compared with its actual brightness at its actual distance $d$, will be the square of the ratio of the distances $(d/10)^2 = R$.

This ratio corresponds to a magnitude difference of $2.5\log_{10}(R)$.

So
$$M = m - 2.5\log_{10}(R)$$

### A specific example: Rigel

Rigel lies at a distance of 237 pc and has an apparent magnitude of 0.12.

When brought to a distance of 10 pc its brightness will increase by the factor $(237/10)^2 = 562$. This corresponds to a magnitude difference of $2.5 \times \log_{10}(562)$, and thus the absolute magnitude of Rigel is given by:

$$M = 0.012 - 2.5 \times \log_{10}(562)$$
$$= 0.12 - (2.5 \times 2.75)$$
$$= -6.7$$

Rigel has an absolute magnitude of $-6.7$.

### Another example: the Sun

The Sun has an apparent magnitude of $-26.75$.

A parsec is $3.08 \times 10^{13}$ km and 1 AU is $1.496 \times 10^8$ km so the Sun lies at a distance of:

$$d = 1.496 \times 10^8 / 3.08 \times 10^{13} \text{ km}$$
$$= 4.86 \times 10^{-6} \text{ pc}$$

When moved to a distance of 10 pc, its brightness will reduce by a factor of $(4.86 \times 10^{-6}/10)^2$ which is equal to $2.36 \times 10^{-13}$. Its absolute magnitude will thus be given by:

$$M = -26.75 - [2.5 \times \log_{10}(2.36 \times 10^{-13})]$$
$$= -26.75 - [2.5 \times (-12.63)]$$

$$= -26.75 + 31.57$$
$$= +4.82$$

The Sun has an absolute magnitude of $+4.82$. (Precise value is $+4.83$)

Given their absolute magnitudes it is possible (and instructive) to compare the luminosity of Rigel with that of the Sun:

The difference in their magnitudes $\Delta m$ is given by:

$$\Delta m = [-6.7 - (+4.82)]$$
$$= 11.52 \text{ magnitudes}$$

The ratio of their brightnesses is given by:

$$R = 2.512^{11.52}$$
$$= 40\,571$$

So Rigel has a luminosity ~41 000 times that of our Sun.

In contrast, the nearest star to us, Proxima Centauri, has an absolute magnitude of $+15.5$. The difference in absolute magnitudes, $\Delta m$, is given by thus

$$\Delta m = 15.5 - 4.82 \text{ magnitudes}$$
$$= 10.68 \text{ magnitudes}$$

The ratio of their brightness is given by:

$$R = 2.512^{10.68}$$
$$= 18\,715$$

So Proxima Centauri has a luminosity which is ~19 000 times less than that of our Sun.

It follows that Rigel is ~45 000 × 19 000 = 855 000 000 times brighter than Proxima Centauri!

You can see that there is a very wide range of luminosity between the brightest stars (like Rigel) and the faintest (like Proxima Centauri). When luminosities are plotted it is thus natural to use a logarithmic scale. The Sun actually turns out to be quite a good star to take as the reference luminosity, as it lies roughly halfway (on a logarithmic scale) between the faintest and brightest stars.

## Colour and surface temperature

If one spends a little time observing the night sky, it soon becomes apparent that stars can have different colours: Betelgeuse in Orion and Aldebaran in Taurus have an orange tint, Capella, in Auriga, is somewhat yellow in colour and Rigel, also in Orion, is a slightly bluish white. Brocchi's Cluster, commonly called 'The Coathanger', is not a true cluster, but a chance arrangement of stars called an 'asterism'. Figure 6.3 shows that star colours can range from red through to white and blue.

The colours we perceive are a function of the surface temperature of the star. As the surface temperature increases from ~3000 up to ~20 000 K, the colour of the star moves from red, through orange, yellow and white to blue. Stars act as approximate black body radiators and, as was described in Chapter 2, both the peak wavelength and total power output of a black body are related to their temperature, so giving us two ways of estimating a star's surface temperature.

Let us calculate the surface temperature of Proxima Centauri. From the calculation above of the luminosity of Proxima Centauri relative to our Sun, and given the Sun's luminosity of ~$4 \times 10^{26}$ W, it has a luminosity given by:

$$L = \text{~}4 \times 10^{26}/19\,000\,\text{W}$$
$$= 2.1 \times 10^{22}\,\text{W}$$

**Figure 6.3** Impression of Brocchi's Cluster. The effect of a 'soft-focus' filter has been used to spread the stellar images so that star colours are more easily seen.

In 2002, the Very Large Telescope used a special technique (to be described later) to measure an angular diameter of $1.02 \pm 0.08$ milliarcsec for Proxima Centauri. Given its distance of ~1.3 pc the actual diameter of Proxima Centauri can be calculated to be about 1/7th that of the Sun. Using the Stephan–Boltzmann Law:

$$L = \sigma A T^4$$
$$= 5.671 \times 10^{-8} \times 4 \times \pi \times (1 \times 10^8)^2 \times T^4$$

So,
$$T = \{2.1 \times 10^{22}/[5.671 \times 10^{-8} \times 4 \times \pi \times (1 \times 10^8)^2]\}^{-4}$$
$$= (2.1 \times 10^{22}/7.1 \times 10^9)^{-4}$$
$$= 1300\,K$$

One can use a somewhat simplified approach relating the temperature to that of the Sun: Proxima Centauri is ~1/7th the diameter of the Sun, thus its surface area, $A$, is ~1/49 times smaller. Its luminosity, derived above, is ~19 000 times less. From the Stephan–Boltzmann Law $L$ is proportional to $AT^4$, so the ratio of its surface temperature, $T_{PC}$, to that of the Sun is given by:

$$T_{PC}/T_{Sun} = [(1/19\,000)/(1/49)]^{-4}$$

So,
$$T_{PC} = 0.22 \times 5800\,K$$
$$= 1300\,K$$

We can use the same method for Rigel.

Rigel is 62 times the diameter of the Sun, thus its surface area is ~3800 times greater. Its luminosity, derived above, is ~45 000 times greater. From the Stephan–Boltzmann Law, $L$ is proportional to $AT^4$, so the ratio of its surface temperature, $T_{Rigel}$, to that of the Sun is given by:

$$T_{Rigel}/T_{Sun} = (45\,000/3800)^{-4}$$

So,
$$T_{Rigel} = 1.85 \times 5800\,K$$
$$= 10\,700\,K$$

This basic idea can make some calculations very easy. Here is an example:

A star has twice the diameter of the Sun and twice its surface temperature. What is its luminosity compared with that of the Sun?

The star's surface area will be $2^2$ or 4 times that of the Sun. Each square metre of its surface will radiate $2^4$ or 16 times more energy. Thus, in total, the star's luminosity will be 64 times that of the Sun.

## 6.6    Stellar photometry

The black body radiation curves in Figure 2.3 show that stars of given temperatures will emit differing amounts of energy at different wavelengths. For example, a star whose surface temperature is 7000 K will emit nearly three times as much energy in blue light than in red light. Conversely, a star at 5000 K will emit about twice as much energy in the red part of the spectrum than in the blue. This enables measurements of the apparent magnitudes at two colours, say blue and red, to be used to give a measure of temperature. Astronomers measure the colour index of a star by subtracting the magnitude in one spectral range from that in another. A standard set of filters has been defined for use at observatories all over the world: the ultraviolet band is called U magnitude, the blue band the B magnitude, the yellow band (which roughly matches the spectral sensitivity of the eye) the V magnitude, and the red band the R magnitude. Thousands of stars have had their magnitudes measured in the U, B and V filters – in so-called UBV photometry. The difference between the magnitudes in the B and V bands is called the colour index. A very hot star is brighter in the blue than in the yellow part of the spectrum, so the V magnitude will be higher and thus the colour index will be negative. The colour index is zero for a star whose temperature is ~10 000 K; −0.3 for the very hottest stars and ~+2 for the coolest.

## 6.7    Stellar spectra

When discussing the Sun, the formation of the Fraunhofer absorption lines observed in its spectrum was described. It turns out that the absorption lines that are seen in the spectra of stars are very closely related to the temperature of the gas through which the light passes on its way from their visible surfaces. If the temperature of a gas was at absolute zero, all the electrons would be in the lowest possible energy levels – the ground states of the atoms. As the temperature rises, first some electrons will rise into their first excited state and then to higher excited states until, at sufficiently high temperatures, electrons may escape from their atoms which are then said to be ionized. The remnant of an atom will then have less than its full quota (one for each proton) of electrons and is called an ion (often called a positive ion as, having lost electrons having negative charge, it will have a net positive charge). Roman numerals are used to define the state of ionization of atoms. For example, H I represents a neutral hydrogen atom and H II an ionized hydrogen atom that has lost its one electron. In the same way He I is neutral helium, He II singly ionized helium and He III doubly ionized helium. The roman numeral is simply the number of lost electrons +1.

### 6.7.1     *The hydrogen spectrum*

Neutral hydrogen produces a distinctive set of spectral lines, called the Balmer series that range in wavelength from 363.46 nm in the ultraviolet to 656.3 nm in the red. The most prominent is the 635.3 nm red line which is called the Hα line and is seen in clouds of gas where the electrons have been lifted into excited states by incident ultraviolet radiation and then drop back down into lower energy states. To form the Balmer series of visible lines the electrons drop down to the first excited state called level 2. The Hα line is caused by a transition from the second to the first excited state and thus from level 3 down to level 2. The green line, produced by the transition from level 4 down to level 2, is called the Hβ line whilst the Hγ line, in the mid blue, is produced by the transition from level 5 down to level 2 (Figure 6.4).

Our eyes are not sensitive to the red light of the Hα line and, sadly, we see very little colour in the universe with our eyes, but our eyes are far more sensitive in the green, and using a telescope of ~16 in. or more aperture some objects, such as the Dumbell Planetary Nebula, appear a vivid green – the light from the Hβ line.

Other atoms will produce similar series of lines either in neutral or ionized form dependant on the temperature of the stellar atmosphere. The absorption spectra that we observe are thus a mix of all these lines and depend strongly on temperature. The hydrogen Balmer series, for example, appears strongest when the star's atmosphere is ~9000 K. At very high stellar temperatures, virtually all the hydrogen atoms are ionized so the Balmer lines are very weak. However, it takes far more energy to fully ionize helium so lines of both neutral (He I) and singly ionized (He II) helium are seen.

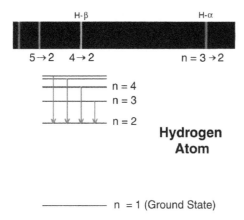

**Figure 6.4** The transitions that give rise to the Balmer series of spectral lines.

## 6.7.2    *Spectral types*

In the latter part of the nineteenth century, the spectra of thousands of stars were photographed by astronomers at Harvard University and the spectra were used to classify the stars into what are called their spectral types. For example, Type A stars were those where the hydrogen Balmer lines were seen to be at their strongest. Stars where the hydrogen lines were weak but helium lines were seen were called Type O. In all, the stars were split into seven spectral types: O, B, A, F, G, K and M. Here they have been listed in decreasing order of temperature, O the hottest and M the coolest. Each type is split into tenths, so the hottest stars within a spectral type will be classified as, say, G0 and the coolest within that type G9. Our Sun is classified as a G2 star and is thus towards the hotter end of the G type stars.

O type stars range from ~60 000 down to 30 000 K. As we will see in Chapter 7, such stars have a very short lifetime so are relatively rare. They are indicated by lines of ionized helium (He II) in their spectra – only possible at very high temperatures.

B type stars are cooler, ranging from 30 000 down to 10 000 K. The hydrogen Balmer lines are stronger, and lines of neutral helium (He I) are seen rather than ionized helium.

A type stars range in temperature from 10 000 down to 7500 K. As previously mentioned, the hydrogen Balmer lines are strongest here and lines of singly ionized elements such as magnesium and calcium appear.

F type stars cover the range from 7500 down to 6000 K. The Balmer lines of hydrogen are weaker whilst those from singly ionized calcium (Ca II) are becoming prominent.

G type stars, the type which includes our Sun, cover the temperature range from 6000 down to 5000 K. (Remember our Sun is a G2 type star with a surface temperature that we calculated above to be 5800 K.) The H and K lines of singly ionized calcium are at their strongest.

K type stars range from 5000 down to 3500 K and the many spectral lines come largely from neutral metals such as iron and sodium.

M type stars are the coolest with surface temperatures less than 3500 K. At such temperatures molecules can exist in the stellar atmosphere and so their spectra show many molecular lines.

Analysis of a star's absorption spectrum is thus an excellent method of determining its temperature to add to those previously discussed (Figure 6.5).

From the surveys that currently exist, the percentages of stars in the differing spectral classes are given in Table 6.1.

From Table 6.1 it can be seen that the great majority of stars are cool M type stars and there are a very small percentage of O and B type stars.

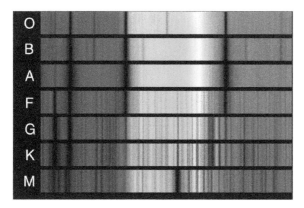

**Figure 6.5**  Typical stellar spectra.

**Table 6.1**  The percentages of stars in the differing spectral classes.

| Type | Colour | Proportion (%) |
|------|--------|----------------|
| O | Blue | 0.003 |
| B | Blue-white | 0.13 |
| A | White | 0.63 |
| F | White-yellow | 3.1 |
| G | Yellow | 8 |
| K | Orange | 13 |
| M | Red | 78 |

## 6.8    Spectroscopic parallax

The fact that the spectra of stars can be measured, even for stars at great distances, allows the stellar distance scale to be extended across the galaxy and even out to the nearest galaxies. The method used is called spectroscopic parallax – though it has nothing to do with parallax! It is based on the very simple premise that all stars of the same spectral type, such as Type F0 star, will have the same intrinsic luminosity. Suppose then that we observed a Type F0 star that appeared 10 000 times less bright than a nearby Type F0 star which was at a distance of 8 pc as measured by the method of parallax. The inverse square law tells us that the distant star would be $(10\,000)^{1/2}$ or 100 times further away, so would lie at a distance of 800 pc.

As a further example, we observe a Type G2 star which has an apparent magnitude of $+9.8$. We know that our Sun has an absolute magnitude of $+4.8$ – which is the apparent magnitude that it would have at a distance of 10 pc. The distant G2 star is thus 5 magnitudes fainter than our Sun would appear if it were to be at a distance of 10 pc. However, 5 magnitudes correspond to a brightness ratio of 100, so the distant star must be $(100)^{1/2}$ or 10 times further away than 10 pc. Hence, the distant G2 star lies at a distance of 100 pc.

As a final example, consider two B8 stars (like Rigel, in Orion). Rigel has an absolute magnitude of $-6.7$, this being the apparent magnitude that it would have at a distance of 10 pc. The distant star lies in a daughter galaxy of our own Milky Way Galaxy called the Large Magellanic Cloud (LMC) and has an apparent magnitude of $+11.7$. The magnitude difference is thus 18.4 magnitudes corresponding to a brightness ratio of:

$$2.512^{18.4} = 23 \times 10^6$$

From the inverse square law it will thus lie at a distance of:

$$10 \times (23 \times 10^6)^{1/2} \text{ pc} = 10 \times 4800 \text{ pc} = 48\,000 \text{ pc}$$

This is thus a measure of the distance of the LMC.

The method of spectroscopic parallax suffers from two fundamental problems. The first is that stars of the same spectral type do not necessarily have the same luminosity as the intrinsic luminosity of a star depends to an extent on what is called the star's metallicity – the percentage of elements heavier than helium or hydrogen in its makeup – so reducing the accuracy of the method. The second is that the correlation between spectral type and luminosity is not perfect: the main sequence is not a thin line, but a band and the scatter in magnitude for a specific spectral type (such as F0) is of order $+/-1$ magnitude. One magnitude is a brightness ratio of 2.512, so this gives a distance error of $+/- (2.512)^{1/2}$ which equals 1.6, so spectroscopic parallax is not that precise.

The measurements of the luminosity of distant stars suffer from a further problem; that of extinction. (This is the name given to the absorption of light by intervening dust.) This would reduce the observed brightness of a star and so make it appear to be further away. There is, however, a way in which this can be detected and corrected for to some extent. The spectral type will indicate that the star has a particular surface temperature. We have also seen how the temperature can be estimated from the colour index. In the absence of extinction these should be approximately the same. However, dust absorbs more light in the blue (shorter wavelength) end of the spectrum than it does at the red (longer wavelength) end of the spectrum. Thus the light from a star whose light has passed through dust clouds will appear redder than it should. This will alter its colour index and indicate a lower surface temperature than that indicated by the spectral index of the star. From this difference it is possible to estimate the effect on the star's apparent brightness and so make a suitable correction.

## 6.9    The Hertzsprung–Russell diagram

In the early 1900s, Ejnar Hertzsprung in Denmark and Henry Russell at Princeton University in the USA independently plotted a graph of stellar luminosity against temperature which has become known as the Hertzsprung–Russell diagram (H–R diagram) (Figure 6.6). Both axes are logarithmic. The *x* axis represents temperature. (Note: it increases to the left of the plot.) Instead of using temperature to define the ordinate, both spectral type and the (B-V) colour index can be used as

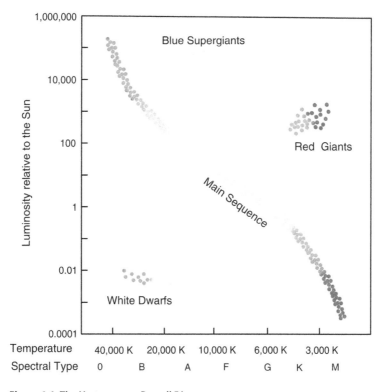

**Figure 6.6** The Hertzsprung–Russell Diagram.

both are directly related to temperature. Henry Russell's first diagram, published in the journal *Nature* in 1914, used the spectral type. An observer might use the B-V index and a theorist would be likely to use temperature, regarding this as being the most fundamental. On the plot shown in Figure 6.6, both the spectral type and the temperature are given.

The vertical axis is a measure of brightness. Here, a theoretician might well use a star's luminosity compared with the Sun, whilst an observer would use absolute magnitude. Russell used the absolute magnitude for this ordinate, but this author prefers the use of luminosity as giving a better understanding of the relative brightness of the stars. When plots are used using the colour index and absolute magnitude, they are often called a colour–magnitude diagram rather than an H–R diagram (Figure 6.7).

To provide a totally true diagram, all stars within a volume of space sufficiently large to include a reasonable number of the very rare O type stars would need to be plotted – this would be known as a complete sample. This is very difficult to achieve as, in this case, the volume is so large that the majority of the faintest stars, those of Type M, would be too faint to be detected, so the plot presented is not a completely true reflection of the relative numbers of the different types of stars. It does, however, show where a representative sample of stars lie on the H–R diagram.

### 6.9.1    *The main sequence*

The main sequence is an S-shaped region that extends from the top left (very bright, high surface temperature O type stars) down to the bottom right (faint, low surface temperature M type stars) of Figure 6.6. It is the region which includes between 80% and 90% of all stars. The stars at the lower right of the main sequence are called red dwarfs as their luminosity is far less than that of our Sun.

### 6.9.2    *The giant region*

Above the main sequence on the right-hand side of Figure 6.6 is an area of bright stars whose colours range through yellow, orange and red. As these stars are very bright they are called giant stars, such as the star Aldebaran, in Taurus, which is called a red giant (though they actually appear orange in colour). At the top of Figure 6.6 is a region, extending from the blue through to the red, of exceedingly bright stars which are called supergiants. Betelgeuse, in Orion, at the extreme top right of the plot is a red supergiant. In contrast, at the extreme top left of the plot is the star Rigel, the brightest star in Orion, whose luminosity we calculated to be ~45 000 times that of the Sun. This is termed a blue supergiant.

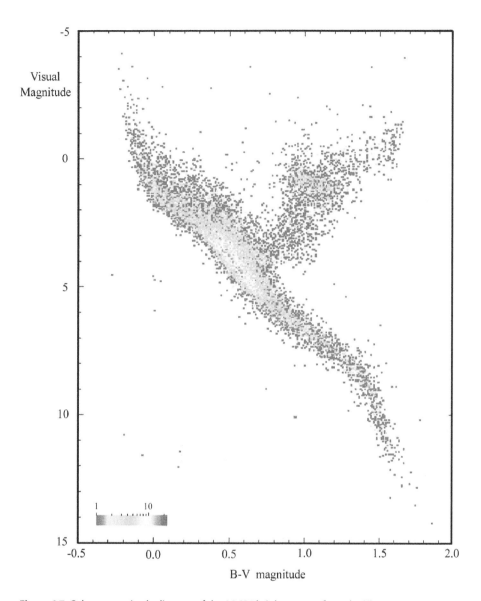

**Figure 6.7** Colour–magnitude diagram of the 16 631 brightest stars from the Hipparcos catalogue showing the number of stars in given colour–magnitude cells. (Note that this greatly overemphasizes the number of bright stars as they can be seen at great distances.) Image: Hipparcos Space Astrometry Mission, ESA.

### 6.9.3    *The white dwarf region*

Below the main sequence lies a region in which white dwarf stars are found. (They encompass a wide surface temperature range and are not necessarily white.) The companion to Sirius, in Canis Major, is a white dwarf. As we will see in Chapter 7 white dwarfs are the remnants of stars like our Sun. They are very small, about the size of the Earth, so even those with very high surface temperatures are not very luminous.

### 6.9.4    *Pressure broadening*

As a red dwarf and a red giant or red supergiant can all have the same surface temperature, one might well ask how they can be distinguished. Obviously their measured luminosity is a major factor but, in addition, their spectra appear different. It is possible to measure the width of spectral lines and it is found that those of the giant stars are less than those of the dwarf stars. As we will shortly see, the outer envelopes of red giant stars are very tenuous with a very low gas pressure whereas the pressure in the atmosphere of M type dwarfs is far higher. The width of an emitted spectral line in gas of higher pressure is broadened in comparison with that in a low pressure due to a phenomenon called pressure broadening: an atom which is able to emit a photon of energy unhindered will produce a narrow band of emitted frequencies but if this atom collides with another atom during the emission process (which will happen far more often at higher pressures), the emitted wave train is shortened in comparison, and this gives rise to a wider range of frequencies – a broader spectral line. This same effect is seen in the absorption lines of stars with red giants having narrower spectral lines than red dwarfs.

The reason why most stars are seen to lie on the main sequence is simply because this is where stars spend the majority of their life as stable objects producing energy by the nuclear fusion of hydrogen to helium. As stars evolve, their position in the H–R diagram changes and a star is said to move along an evolutionary track across the H–R diagram. The tracks of stars of different stellar types will be discussed in Chapter 7. The main sequence is not a line but, as shown, is a band across the H–R diagram. This is because, as they age, stars become somewhat more luminous with an increased surface temperature. They thus move somewhat up and to the left of the diagram during their hydrogen burning phase. The giant phase in the life of a star is relatively brief which is why we see far fewer stars of this type. The white dwarfs are the final states of many stars and gradually cool over billions of years thus moving down and to the right of the H–R diagram. Over time, as more of the stars in our Galaxy come to the end of their lives, their numbers will increase relative to those on the main sequence.

# 6.10　The size of stars

### 6.10.1　*Direct measurement*

The angular sizes of relatively nearby stars can be measured directly. The diameter of our Sun, calculated earlier, comes from knowledge of its angular size and distance. There is only one star, Betelgeuse (a red supergiant in Orion), whose angular size can be directly observed with a normal telescope. In 1995, the Faint Object Camera of the Hubble Space Telescope (HST) was used to capture the first conventional telescope image of Betelgeuse and measured an angular size of ~0.05 arcsec (Figure 6.8). (Betelgeuse has a very diffuse outer envelope so it is rather hard to estimate the angular size.) The distance to Betelgeuse is not precisely known but, if it is assumed to be 131 pc, then one can calculate the diameter:

$$d = D\theta$$

**Figure 6.8** Hubble Space Telescope image of Betelgeuse. Image: A. Dupree (CfA), NASA, ESA.

With θ in radians and *d* (the diameter) and *D* (the distance) in kilometres we get:

$$d = 131 \times 3.1 \times 10^{13} \times [0.05/(3600 \times 57.3)]\,km$$
$$= 4.1 \times 10^{15} \times 2.4 \times 10^{-7}\,km$$

So the diameter is ~1 billion km, about 700 times that of the Sun.

A method called optical interferometry uses two or more mirrors separated by distances of order tens of metres and combines their light to give the effect of one giant optical telescope so giving far higher resolution than single telescopes such as the HST which has a mirror of 2.4 m diameter. They have been used to measure the angular diameters of nearby stars, so that, given their distances as measured by the method of parallax, their diameters can be derived (Figure 6.9).

In 2002, the light from two 8.2-m telescopes of the Very Large Telescope (VLT) array in Chile was combined to form an interferometer with a baseline of 102.4 m. Its resolution was thus equivalent to an optical telescope ~100 m in diameter. It measured an angular diameter of Proxima Centauri, the nearest star to the Earth,

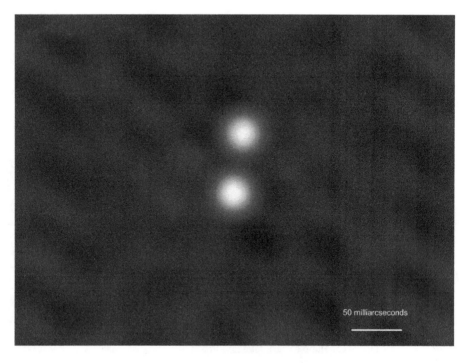

50 milliarcseconds

**Figure 6.9** An image of the double star Capella made with the COAST optical interferometer at the Mullard Radio Astronomy Observatory at Cambridge.

of 1.02 +/− 0.08 thousandth's of an arcsecond – incidentally about the angular size of an astronaut on the surface of the Moon as seen from the Earth!

Proxima Centauri is a red dwarf star that lies at a distance of 1.3 pc so its angular diameter is given by:

$$d = D\theta \text{ (where } \theta \text{ is in radians)}$$
$$= 1.3 \times 3.1 \times 10^{16} \times 1.02/(1000 \times 3600 \times 57.3) \text{ m}$$
$$= 2 \times 10^{8} \text{ m}$$

The Sun has a diameter of $1.4 \times 10^{9}$ m, so Proxima Centauri has a diameter ~1/7th that of the Sun.

## 6.10.2    *Using binary star systems to calculate stellar sizes*

Over 80% of all stars are in binary or multiple star systems. If the plane of the orbit of a binary pair is close to the line of sight, each star will occult the other in turn and brightness of the system will be seen to drop. Such a system is called an eclipsing binary. If the orbital parameters of the system are known, the time during which the larger star is occulted by the smaller gives a measure of the diameter of the occulting star.

The most famous, and the first to be discovered, eclipsing binary is the star system Beta Persei, Algol. It was called the 'demon star' as it appears to 'wink' – its brightness drops by 30% (from magnitude 2.1 down to 3.4) for a total of ~10 h precisely every 2.86739 days or ~68.8 h (Figure 6.10).

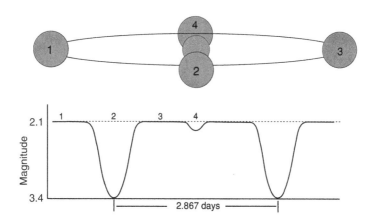

**Figure 6.10** The light curve of the eclipsing binary star Algol.

The two stars are separated by 0.062 AU or 9.275 million km. These facts allow us to estimate the diameter of the occulting star. The circumference of the orbit is 29.1 million km, $(\pi \times D)$, so if the eclipse lasts 10 h, which is 14.5% of the 68.8 h period, the diameter will be approximately $(29.1 \times 14.5)/100$. This equals 4.2 million km. Our Sun has a diameter of 1.4 million km so the occulting star has a diameter ~3 times that of our Sun.

### 6.10.3  *Using the Stephan–Boltzman Law to estimate stellar sizes*

Given a star's position in the H–R diagram, one can use the Stephan–Boltzman Law to calculate its radius. The luminosity of a star compared with our Sun is proportional to the square of the star's diameter relative to the Sun and to the fourth power of its temperature relative to the Sun's temperature.

Rigel has an absolute magnitude of −6.7, our Sun +4.83, a difference of 11.53 magnitudes corresponding to a luminosity difference of:

$$2.512^{11.53} = 41\,000$$

Rigel has a surface temperature of 10 700 K, or 1.84 times that of the Sun. Each square metre of Rigel's surface will thus radiate $1.84^4$ or 11.5 times that of the Sun. Its surface area must thus be 41 000/11.5 or ~3500 times that of the Sun, so its diameter will be $3500^{1/2}$, that is ~59 times that of the Sun.

We can also use this method to derive a diameter of Betelgeuse. It is ~60 000 times brighter than our Sun and its surface temperature is 3500 K which is 0.6 that of our Sun. Each square metre of Betelgeuse's surface will thus radiate $0.6^4$ or 0.13 times that of the Sun. Its surface area must thus be 60 000/0.13 or 460 000 times that of the Sun so its diameter will be $460\,000^{1/2}$ or ~679 times that of the Sun. This agrees quite well with the direct measurement of ~700 times that of our Sun derived earlier.

The star with the brightest apparent magnitude in the northern hemisphere is Sirius A, which is in a binary system with a white dwarf star Sirius B (Figure 6.11). Sirius A is ~26 times brighter than our Sun and Sirius B is 416 times *less* bright than our Sun. Sirius A has a surface temperature of 9900 K, whilst Sirius B has a surface temperature of 15 000 K. Each square metre of Sirius A's surface will thus radiate $1.72^4$ or 8.7 times that of the Sun. Its surface area must thus be 26/8.7 or 3 times that of the Sun so its diameter will be $3^{1/2}$ or ~1.7 times that of the Sun. In contrast, each square metre of Sirius B's surface will radiate $(15\,000/5800)^4$ or 45 times that of the Sun. Its surface area must thus be $(1/416)/45$ or $5.3 \times 10^{-5}$ times that of the Sun so its diameter will be $(5.3 \times 10^{-5})^{1/2}$ or 0.007 times that of the Sun. White dwarf stars are very small and of comparable size with our Earth whose diameter is 0.009 times that of the Sun!

**Figure 6.11** Hubble Space Telescope image of Sirius and its companion white dwarf star, Sirius B (lower left). Image: NASA, H.E. Bond, E. Nelan, M. Barstow, M. Burleigh, J.B. Holberg, STScI, U. Leicester, U. Ariz.

(Estimates of the surface temperature of Sirius B vary rather widely. Due to the 4th power dependence on temperature, this has a major effect on the calculated diameter.)

In general, the radii of stars on the main sequence range from ~20 times that of the Sun at the upper left of the H–R diagram down to 0.1 that of the Sun at the lower right. Giant stars lie in the region of ~10–100 times that of the Sun; an example is Aldebaran, in Taurus, which has a radius 45 times that of the Sun. Supergiant stars such as Betelgeuse often pulsate, its radius varying between around 700–1000 times that of the Sun with a period of ~2100 days. As we have seen, in complete contrast, the diameters of white dwarfs are comparable with that of the Earth, not the Sun (Figure 6.12).

## 6.11    The masses and densities of stars

The mass of our Sun provides the starting point for determining that of other stars. We can determine the mass of a second star if we can observe one in a binary system orbiting around a G2 star like our Sun. The method uses the generalization of Kepler's Third Law that was derived by Isaac Newton. Newton's derivation showed that the square of the orbital period is inversely proportional to the sum of the two masses orbiting each other. This effect had not been included by

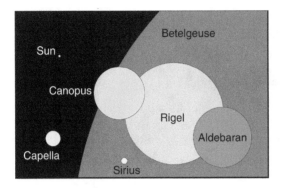

**Figure 6.12 The relative sizes of some typical stars.**

Kepler in his empirical third law as the masses of the planets were very small relative to that of the Sun, but it does affect calculations when one mass is comparable with the other as in the case of a binary star system.

$$P^2 = [4\pi^2/G(M_1 + M_2)]\, a^3$$

So, if the period and orbital major axis of a binary system can be measured, then the sum of the masses can be found. Suppose we observe a binary system in which one star was a Type G2 (assumed to have the same mass as our Sun) and the second a Type A0 star. If the combined mass of the system was calculated to be 4 solar masses, then the mass of the other star would be 3 solar masses. We thus know that an A0 type star would have a mass of 3 solar masses. If then, a binary star system was observed to have a mass of 4.5 solar masses one of which was a Type A0 star, we could deduce that the other star in the system would have a mass of 1.5 solar masses. From its temperature of ~7000 K and spectra we would then know that a star of Type F2 would have a mass of 1.5 solar masses.

From such observations it has been found that the most massive stars are about 50 times more massive than the Sun with the least massive being about 1/15th the mass of the Sun.

## 6.12    The stellar mass–luminosity relationship

Stars on the main sequence have been found to have a well defined mass–luminosity relationship. The luminosity rises steeply with mass. This is the result of the fact that, in order to support the mass of the overlying star, the pressures (and hence temperatures) of the star's core have to increase as the star's mass increases (Figure 6.13).

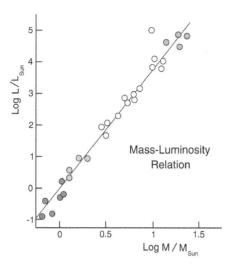

**Figure 6.13** The mass–luminosity relationship.

## Stellar lifetimes

The nuclear fusion rate increases rapidly with temperature so that more massive stars burn up their hydrogen fuel far more rapidly than less massive stars. Assuming that stars can burn similar percentages of their total mass in the core, then a natural consequence is that more massive stars will have shorter lives on the main sequence.

Let us take Rigel as an example: its mass is 17 times that of our Sun whilst its luminosity is 41 000 times that of our Sun. It will thus only stay on the main sequence for 17/41 000 of the lifetime of our Sun. We believe that the Sun will remain on the main sequence for ~10 billion years which implies that Rigel can only remain there for ~1/2600 of this period – about 4 million years! This is actually an overestimate as Rigel, being very hot, emits a good percentage of its radiation in the ultraviolet which is not accounted for in the absolute *visual* magnitude. There is a second measure of luminosity called the bolometric luminosity which measures the total energy output across the whole of the electromagnetic spectrum and this is ~66 000 times that of our Sun rather than ~41 000 so reducing the main sequence lifetime to ~2.7 million years.

At the other end of the main sequence, a red dwarf star might have a luminosity of ~1/10 000 that of our Sun, but a mass of 1/5th that of the Sun. This would give it a lifetime of ~2000 times that of the Sun and thus the lifetime is far longer than the present age of the universe. In fact, whereas in a star like the Sun only about 10% of the star's mass will be converted into helium during its main sequence phase, the less massive red dwarfs mix their interiors by convection so allowing a greater proportion of their mass to be converted into helium and thus extending their lifetimes even further.

## 6.14    Questions

1. A star has an apparent magnitude of 17 and a measured parallax of 0.2 arcsec. Calculate its absolute magnitude.
2. A star at a distance of 100 pc has an apparent magnitude of 15. What is its absolute magnitude?
3. A distant F0 star is 6.3 times less bright than a nearer F0 star which has a parallax of 0.2 arcsec. Calculate its distance in parsecs.
4. The star Procyon has an absolute magnitude of 2.65. The Sun has an absolute magnitude of +4.82. How much brighter is Procyon than the Sun?
5. A star has 3 times the diameter of our Sun, and 2.5 times its surface temperature. What is its luminosity compared with that of our Sun?
6. A star has a radius and surface temperature that are both twice that of our Sun. What will be its radiated energy output compared with that of our Sun?
7. The star Achenar is 6 times more massive than our Sun. Use the mass–luminosity function (Figure 6.13) to estimate its luminosity relative to our Sun. (Remember this is a logarithmic plot!)
8. The star Dubhe in Ursa Major is 300 times more luminous than our Sun. Use the mass–luminosity function (Figure 6.13) to estimate its mass relative to our Sun. (Remember this is a logarithmic plot!)
9. A star has a mass which is 20 times that of the Sun and a surface temperature of 30 000 K. Given that the Sun has a surface temperature of 6000 K and assuming both stars have a similar density, estimate the star's lifetime on the main sequence given that our Sun will remain on the main sequence for 10 000 million years. (Assume both stars act as black bodies and convert a similar percentage of their mass into energy.)
10. A star has a surface temperature twice that of our Sun. Calculate how much energy each square metre of it surface will radiate compared with that of the Sun. The star has a mass 3 times that of our Sun and has a similar density. Calculate its radius and hence surface area compared with that of the Sun. Hence, calculate how much more energy it radiates relative to the Sun and estimate its lifetime compared with the Sun.

## Chapter 7

# Stellar Evolution – The Life and Death of Stars

We have seen how stars are formed and how our Sun burns hydrogen to helium whilst lying on the main sequence. This chapter will look at how stars in different mass ranges evolve, both on the main sequence and during the later stages of their life, and describe the remnants left when they die: white dwarfs, neutron stars and black holes. The distinction between low, mid, and high mass stars seem somewhat arbitrary in the literature, so the mass ranges have been specified in the headings to follow.

**7.1**  ## Low mass stars: 0.05–0.5 solar masses

For a collapsing mass of gas to become a star, nuclear fusion has to initiate in its core. This requires a temperature of ~10 million K and this can only be reached when the contracting mass is greater than ~$10^{29}$ kg, about 1/12 the mass of the Sun, or 80 times that of Jupiter.

In low mass stars the conversion of hydrogen to helium by nuclear fusion is the same as in our Sun. However, in stars of greater mass nuclear fusion only converts ~10% of the mass of the star (that residing in its core), whilst in the lowest mass stars it is thought that convection currents that mix the star's interior will allow much of the star's mass to undergo nuclear fusion so increasing its lifetime on the main sequence – a period which is significantly longer than the present age of the universe. We thus have no direct observational evidence of what happens when fusion ceases in such stars and can only use computer modelling to investigate what might happen.

We will see that in order for helium to fuse into heavier elements, temperatures of order 100 million K are required, and this requires sufficient mass in the star's envelope to provide the required pressure to enable such temperatures to be reached. In stars of mass less than 0.5 solar masses there is simply not enough pressure to give the temperatures that would allow helium fusion to begin. Consequently, when nuclear fusion, converting hydrogen to helium, finally ceases – and modelling of a

*Introduction to Astronomy and Cosmology*  Ian Morison
© 2008 John Wiley & Sons, Ltd

0.1 solar mass red dwarf suggests that this might be after 6 trillion years – the star will slowly collapse over a period of several hundred billion years to form what is called a white dwarf. Over many trillions of years, the white dwarf will cool until its surface temperature is below that at which significant light is emitted and the inert remnant will become a black dwarf. No white dwarfs derived from low mass stars yet exist, but they will be discussed in detail in the next section as their evolution also produces white dwarfs which *can* now be observed.

## 7.2    Mid mass stars: 0.5–~8 solar masses

All stars in this range will have a common end state in the form of a white dwarf. There is, however, a difference in the process of nuclear fusion from hydrogen to helium in stars above and below ~2 solar masses.

For stars with mass less than ~2 solar masses, like our Sun, the bulk of their energy is produced by the proton–proton cycle described in Chapter 2. However, there is a more complex process called the carbon–nitrogen–oxygen (CNO) cycle that provides 1–2% of the Sun's total energy output (Figure 7.1). In stars

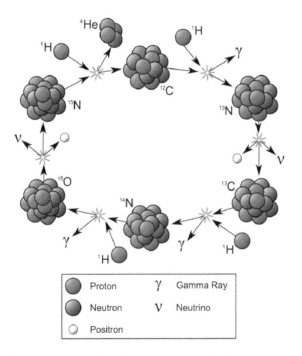

**Figure 7.1** The carbon–nitrogen–oxygen cycle. Image: Wikipedia Commons.

with mass greater than ~2 solar masses this process, proposed independently by Carl von Weizsäcker and Hans Bethe in 1938 and 1939, respectively, becomes dominant. It provides a very efficient way of converting hydrogen to helium so the hydrogen in more massive stars burns more quickly so increasing the energy output of the core. As the greater energy output of the stars must be balanced by radiation from their surface, the stars become bluer and have a greater luminosity – as shown by their positions on the Hertzsprung–Russell diagram (H–R diagram).

The reactions of the CNO cycle are:

$$^{12}C + {}^{1}H \rightarrow {}^{13}N + \gamma$$
$$^{13}N \rightarrow {}^{13}C + e^{+} + \nu_{e}$$
$$^{13}C + {}^{1}H \rightarrow {}^{14}N + \gamma$$
$$^{14}N + {}^{1}H \rightarrow {}^{15}O + \gamma$$
$$^{15}O \rightarrow {}^{15}N + e^{+} + \nu_{e}$$
$$^{15}N + {}^{1}H \rightarrow {}^{12}C + {}^{4}He$$

The net result of the cycle is to fuse four protons into an alpha particle along with two positrons, two electron neutrinos (which carry some energy away from the star) and three gamma rays. The carbon acts as a catalyst and is regenerated. The two positrons annihilate electrons releasing energy in the form of gamma rays; each annihilation gives rise to two gamma rays.

When the CNO process reaches an equilibrium state, the reactions of each stage will proceed at the same rate. The slowest reaction within the cycle is that which converts $^{14}N$ into $^{15}O$ so, in order for this reaction to have an equal reaction rate, the number of nitrogen nuclei must be significantly larger than that of carbon or oxygen. Thus, over time, the relative amount of nitrogen increases until equilibrium is established. Whilst the total number of the carbon, nitrogen and oxygen nuclei is conserved, nitrogen nuclei become the most numerous, regardless of the initial composition. This process produces essentially all of the nitrogen in the universe and thus has great significance for us as nitrogen is an essential element of all life-forms here on Earth.

## 7.2.1    Moving up the main sequence

As the proton–proton or CNO cycles convert hydrogen to helium in the core of the star, its mean molecular weight increases. The pressure of an ideal gas increases with the temperature of the gas, but decreases as the mean molecular weight of the gas rises. Assuming (correctly) that the matter in the core acts like an ideal gas, in order for the pressure to remain sufficient to support the overlying layers of the star as the amount of helium increases, the core's temperature must rise. The rate of, for example, the proton–proton chain is a function of the fourth power of the temperature, so that even though the percentage of hydrogen is reducing, the conversion rate of hydrogen to helium gradually increases whilst the star lies on the main sequence. This, of course, increases the energy output of the star, so its luminosity will increase and, in order to radiate more energy, so will its surface temperature. The combined result is that the star slowly moves up and to the left of the main sequence.

It is believed that, when formed, our Sun was ~30% less bright than at the present time and that over the next billion years its luminosity will increase by a further 10%. An obvious consequence is that our Earth will eventually become too hot for life (as we know it) to exist on its surface.

### 7.2.2    *The triple alpha process*

Eventually, either by the proton–proton chain or the CNO cycle, the core of the star will be converted into $^4$He. At this point nuclear fusion stops so that the pressure in the core that prevents gravitational collapse drops. The core thus reduces in size but, as it does so, its temperature will rise. Finally, when the temperature reaches ~100 million K, a new reaction, known as the triple alpha process ($3\alpha$), occurs. It is so-called because it involves three helium nuclei which are also known as alpha particles (Figure 7.2). This is an extremely subtle process. The first obvious nuclear reaction that would occur in a core composed of helium is that two $^4$He nuclei fuse to form $^8$Be. However, $^8$Be is very unstable – it has a lifetime of only $10^{-19}$ s – and virtually instantly decays into two $^4$He nuclei again. Only when the core temperature has increased to 100 million K, does it become likely that a further $^4$He nucleus can fuse with $^8$Be before it decays. The result is

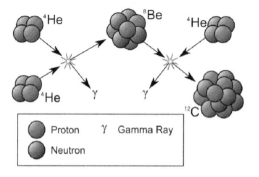

**Figure 7.2** The triple alpha process. Image: Wikipedia Commons.

a $^{12}$C nucleus. It is highly significant to our existence here on Earth that there is such a difference in temperature between that (~15 million K) at which the hydrogen fuses to helium and that (~100 million K) at which $^{12}$C can be formed. If this were not the case, and the process could happen at the core temperatures close to that at which the proton–proton or CNO cycles operate, there would be no long period of stability whilst the star remains on the main sequence with a relatively constant luminosity. This, of course, has allowed stable temperatures to exist on Earth for billions of years and so has enabled intelligent life to evolve.

There is a further real problem in attempting to form $^{12}$C. A temperature of 100 million K is required to give the $^{4}$He nuclei a reasonable chance to fuse with a $^{8}$Be nucleus before it has a chance to decay. The $^{4}$He nuclei are thus moving very fast and so have appreciable kinetic energy. It would be expected that this energy would prevent a stable $^{12}$C nucleus arising as it would be sufficient to split the newly formed nucleus apart. [If a white billiard ball ($^{4}$He) approached a red ball ($^{8}$Be) very slowly they might just 'kiss' and remain touching, but if it came in at high speed the energy of impact would split them apart.]

So why is $^{12}$C so common? This problem was pursued with great vigour by the British astrophysicist, Fred Hoyle, in the early 1950s. As he then stated: 'Since we are surrounded by carbon in the natural world and we ourselves are carbon-based life, the stars must have discovered a highly effective way of making it, and I am going to look for it.'

He realised that the excess energy that was present in the reaction (and thus expected to break up the newly formed $^{12}$C nucleus) could be contained if there happened to be an excited state (called a 'resonance' by particle physicists) of the carbon nucleus at just the right energy above its ground state. This is because, due to the quantum nature of matter, though atomic nuclei usually exist in their ground state, it is possible for them to absorb energy (such as an interaction with a gamma-ray photon) and jump into an excited state. This will later decay back to the ground state with the emission of a gamma ray of the same energy. This is analogous to an atom absorbing a photon of energy which lifts an electron to a higher energy level. The electron will then, in one or more steps, drop back down the energy levels emitting photons as it does so.

Hoyle realised that a stable carbon nucleus could only result if it had an excited state that was very close in energy to that of the excess energy of the three $^{4}$He that came together in its formation. This would thus lift the resulting $^{12}$C nucleus into an excited state from which it could drop back to the ground state by the emission of a gamma-ray photon and so reach a stable state.

Some experiments in the late 1940s had suggested that such an excited state might exist, but Hoyle had been told that these were in error. Hoyle argued that there *must* be an appropriate excited state otherwise we could not exist and pestered the particle physicists at the California Institute of Technology (Caltech), led by William Fowler, to repeat the experiments. Fowler did so (it is said, only so that Hoyle would go away) and found that there was indeed an excited state within 5% of the energy predicted by Hoyle! Hoyle was essentially using the 'anthropic principle', which states that our existence as observers puts constraints on the universe in which we live. William Fowler received the Nobel Prize in part for this work. Many believe that Hoyle should also have won the Nobel Prize for his incisive observation and his following work in showing how the elements are synthesized in stars.

### 7.2.3    *The helium flash*

The onset of helium burning in stars of this mass range is very dramatic. What happens is related to the properties of the electrons that also exist in the core. Under the extreme densities within the core the electrons cannot be regarded as an ideal gas, but begin to show quantum-mechanical aspects in their behaviour. They begin to act as what is termed a degenerate electron gas and behave differently. In an ideal gas, as the temperature increases, so will the pressure and thus the gas will tend to expand so reducing the temperature. This tends to prevent a rapid increase in temperature. However, in a degenerate gas the pressure does not rise, so the temperature in the centre of the star increases rapidly and, as the reaction rate is highly dependent on temperature, the reaction can 'run away' and produce an explosive release of energy in a very short period – perhaps as little as 30 s. This is called the helium flash. However, due to the overlying layers of the star this energy will take thousands of years to reach the surface.

For a given mass of gas, the $3\alpha$ process only releases about 10% of the energy produced in forming helium nuclei from hydrogen. Hence, the length of the helium burning phase will be about 10% of the star's life on the main sequence.

During its helium burning phase the core will be compressed to perhaps 1/50th of its original size and have a temperature of ~100 million K with, in addition, a shell of hydrogen burning surrounding the core. The energy so produced, in part by the shell of hydrogen burning, causes the outer parts of the star to also undergo significant changes. The radius of the star as a whole increases by a factor of ~10, but at the same time the surface cools to (in the case of a 1 solar mass star) a temperature of ~3500 K. The star will then have an orange colour and the star becomes what is called (perhaps perversely) a 'red giant'. The result is that the star's position moves up and to the right of the H–R diagram into the red giant region.

For midmass stars less in mass than our Sun, this is about as far as nuclear fusion can take the formation of elements as there is not enough overlying mass above the core to allow its temperature to rise sufficiently for further nuclear fusion reactions to be carried out.

The stars in the upper part of this mass range are able to carry out one further nuclear reaction:

$$^{12}C + {}^4He \rightarrow {}^{16}O + \gamma$$

This reaction and the $3\alpha$ process are thought to be the main source of carbon and oxygen in the universe today. However, we could not exist if these elements stayed within the stars. They *must* lose much of their material into space (this is the anthropic principle again); this is exactly what is observed.

## 7.3    Variable stars

In the latter stages of its life, a star become less stable and may well oscillate in size. As the star's size increases, its surface area will also increase, tending to increase its luminosity but, at the same time, the surface temperature will reduce so reddening its colour. As the emitted energy per unit area decreases as the fourth power of the temperature, the star's luminosity actually falls as the size increases. Conversely, as the size of the star reduces, the colour will shift towards the blue and the luminosity will increase. The periodic changes in colour and luminosity result in what is called a 'variable star' (Figure 7.3).

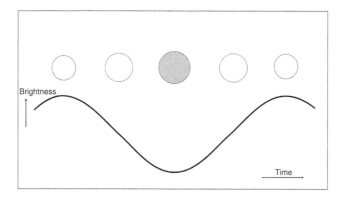

**Figure 7.3** The changes in colour and brightness of a variable star.

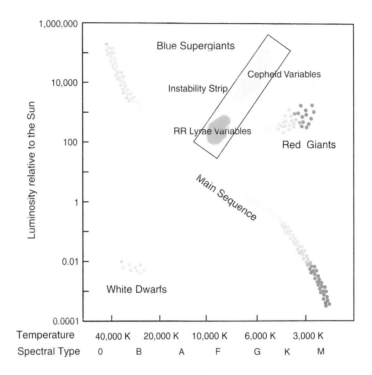

**Figure 7.4** The region where variable stars are found in the H–R diagram.

Variable stars lie in a part of the H–R diagram known as the instability strip, which lies in the upper half of the diagram (Figure 7.4). During this phase of their life, stars often have intense solar winds and so lose much of their outer envelopes into space.

Types of variable stars are named after a 'prototype star', usually the first of a particular type of variable star to be discovered. The RR Lyrae variable stars shown in Figure 7.4 are thus named after the star RR Lyrae. They have masses about half that of our Sun, pulsate with periods of 0.2–2 days and are in the phase of their life when they are burning helium into carbon. They are typically 40–50 times brighter than our Sun, and their brightness is related to the period with which they vary in brightness. This gives them the status of a standard candle. Suppose we have found by parallax the distance of a nearby RR Lyrae variable which has a period of 1 day. If we then observed a distant RR Lyrae star, having the same period but appearing 10 000 time fainter than the nearby one, we could assume that they had the same intrinsic brightness and so, using the inverse square law, the more distant star would be 100 times further away.

Another type of very bright variable stars, called Cepheid variables, which lie higher in the H–R diagram and are some of the brightest stars that we can observe, have played a very important role in determining the size of the universe and will be discussed in detail in Chapter 8.

## 7.4   Planetary nebula

Finally it appears that the star becomes so unstable that the outer parts of the star are blown off to form what is called a planetary nebula surrounding the core remnant. Planetary nebulae are some of the most beautiful objects that we observe in the universe and many, such as the Ring and Dumbbell Nebulae, may be observed with a small telescope (Figure 7.5). Planetary nebulae are relatively common with over 1500 known, but it is expected that many more, perhaps over 50 000, will exist in the Galaxy but are hidden by the dust lanes in our Galaxy. The name planetary nebulae is, of course, a misnomer as they have nothing to do with planets, but many do have a disc-like appearance. They are large tenuous shells of gas which are expanding outwards at velocities of a few tens of kilometres per second. They also contain some dust and have masses of typically one-tenth to one-fifth of a solar mass. About 10 planetary nebulae are thought to be formed each year so the interstellar medium is being enriched by around 1 solar mass per year.

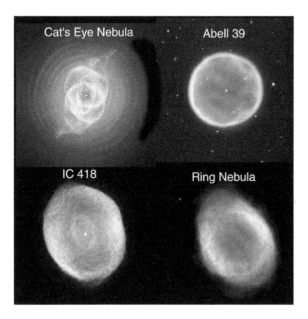

**Figure 7.5** Planetary nebulae. In all cases, the stellar remnant can be seen at the centre.
Image: IC 418, Cat's Eye and Ring Nebulae, Hubble Space Telescope, NASA; Abell 39, NOAO, NSF.

## 7.5    White dwarfs

At the centre of a planetary nebula lies a white or blue-white star. They are not very bright so that relatively large telescopes are required to see them visually. (The author has once, using a 16 in. telescope under perfect conditions, observed the star at the centre of the Ring Nebula.) This star is approaching the final stage of its life when it will become a 'white dwarf'. Once nuclear reactions have ceased, what is left at the centre of the star will contract under gravity. It is composed mainly of carbon and oxygen, and devoid of its outer layers through a combination of the intense stellar winds and the ejection of a planetary nebula. The fact that contraction finally ceases is due to a quantum-mechanical effect known as degeneracy pressure. In 1926, R.H. Fowler realised that, as a result of the Pauli exclusion principle, no more than two electrons could occupy a given energy state. As the allowed energy levels fill up, the electrons begin to provide a pressure – the electron degeneracy pressure – which finally halts the contraction. This pressure only depends on density, not temperature, and this has the interesting result that the *greater* the mass of the white dwarf, the *less* its radius!

A further consequence of being supported by electron degeneracy pressure is that there is a limiting mass which cannot be exceeded. This depends on the composition of the star; for a mix of carbon and oxygen, it turns out to be ~.4 solar masses. This result was published in 1931 by Subramanyan Chandrasekhar when he was only 19 years old! (To be totally accurate, Chandrasekhar had assumed that there would be a greater percentage of heavier elements within the white dwarf so resulting in a limit of 0.91 solar masses.) In 1983, Chandrasekhar rightly received the Nobel Prize for this and other work. We will see later what happens when the mass of the collapsing stellar remnant exceeds the Chandrasekhar Limit.

White dwarfs range in size from 0.008 up to 0.02 times the radius of the Sun. The largest (and thus *least* massive) being comparable with the size of our Earth whose radius is 0.009 times that of the Sun. The masses of observed white dwarfs lie in the range 0.17 up to 1.33 solar masses so it is thus obvious that they must have a very high density. As a mass comparable with our Sun is packed into a volume 1 million times less, its density must be of order 1 million time greater – about 1 million grams per cubic centimetre. (A ton of white dwarf material could fit into a matchbox!)

### 7.5.1    *The discovery of white dwarfs*

The first known white dwarf was discovered by William Herschel in 1783; it was part of the triple star system 40 Eridani. What appeared surprising was that although its colour was white (which is normally indicative of bright stars) it had a very low luminosity. This is, of course, due to its small size so, although each square metre is highly luminous, there are far fewer square metres!

The second white dwarf to be discovered is called Sirius B, the companion to Sirius, the brightest star in the northern hemisphere. Friedrich Bessel made very accurate measurements of the position of Sirius as its proper motion carried it across the sky. The motion was not linear and Bessel was able to deduce that Sirius had a companion. Their combined centre of mass *would* have a straight path across the sky but both Sirius and its companion would orbit the centre of mass thus giving Sirius its wiggly path. Due to its close proximity with Sirius, Sirius B is exceedingly difficult to observe as it is usually obscured by light scattered from Sirius within the telescope optics. A very clean refractor has the least light scatter, and it was when Alvin Clark was testing a new 18 in. refracting telescope in 1862 that Sirius B was first observed visually.

### 7.5.2    *The future of white dwarfs*

The observed surface temperatures of white dwarf stars range from 4000 up to 150 000 K so they can range from orange to blue-white in colour. Their radiation can only come from stored heat unless matter is accreting onto it from a companion star. As their surface area is so small it takes a very long time for them to cool; the surface temperature reduces, the colour reddens and their luminosity decreases. The less the surface temperature the less the rate of energy loss, so a white dwarf will take a similar time to cool from 20 000 down to 5000 K as it will from 5000 to 4000 K. In fact, the universe is not old enough for any white dwarfs to have cooled much below 4000 K; the coolest observed so far, WD 0346+246, has a surface temperature of 3900 K.

### 7.5.3    *Black dwarfs*

Eventually the white dwarf will cool sufficiently so that there is no visible radiation and will then become a black dwarf. They could still, however, be detected in the infrared, though will be very faint, and the presence of those in orbit around normal stars could still be deduced by the effect they have on the motion of a companion star.

## 7.6    The evolution of a sun-like star

Figure 7.6 illustrates the sequence of events in the evolution of a 1 solar mass star, like our Sun, as it 'moves' across the H–R diagram in what is called its evolutionary track. The numbers shown on the evolutionary track in Figure 7.6 correspond to:

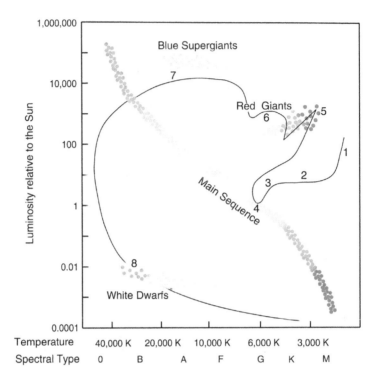

**Figure 7.6** The evolutionary track of a 1 solar mass star on the H–R diagram.

(1)    The surface of the collapsing protostar reaches a temperature when it begins to emit red light. As it is very large, it is initially relatively bright, but drops in brightness as its surface area reduces.

(2)    For a period, the protostar remains at ~10 times the luminosity of the resulting star. The star is getting smaller, which reduces the surface area but, at the same time, the surface temperature is rising so increasing the emission per unit area. The two effects roughly cancel, so the protostar remains at a roughly constant luminosity.

(3)    As the star drops onto the main sequence, it goes through a very violent stage when approximately half its mass is lost. It is called the Tau Tauri phase in the life of the star. (Tau Tauri is a triple star system in Taurus where the primary star is dropping onto the main sequence.)

(4)    The star settles on the main sequence burning hydrogen into helium. As the amount of helium in the core increases, its temperature rises and the star becomes somewhat hotter and so moves upward and to the left of the H–R diagram.

(5)    The star becomes a red giant as it burns helium to carbon.

(6)    It then burns carbon into oxygen within the core and becomes unstable, changing in luminosity and colour as it does so.

(7)    Finally, when no more energy can be extracted from the core, the star blows off its outer envelope, giving rise to a planetary nebula.

(8)    The cooling ember of the core, collapsed down to about the size of the Earth, drops to the lower left of the H–R diagram and becomes a white dwarf. It is on the left of the diagram as it will still be very hot but, as it is very small, it will not be very luminous so lies at the bottom of the diagram.

## 7.7    Evolution in close binary systems – the Algol paradox

For much of their life, stars in close binary systems evolve just as described above, but as one or other of the stars ends its main sequence life things get a little strange. Consider Algol, the demon star, which was discussed in Chapter 6 as an eclipsing binary system. It consists of a 3.7 solar mass main sequence star in close orbit with a 0.8 solar mass sub-giant star. This is odd. Both stars were presumably born at the same time so we would expect the more massive star to leave the main sequence first, as more massive stars burn up their fuel more rapidly. This paradox is resolved by the phenomenon of mass transfer between the two stars. The higher mass star of the original binary pair did, indeed, evolve more rapidly, but as it expanded to form a red giant, the outer parts of its envelope began to spill onto its less massive companion as the gas succumbed to its gravitational pull. It thus increases its mass at the expense of the giant star. The sub-giant in the Algol system *was* originally the more massive of the two stars but has lost a lot of mass to its companion.

The future of this system is equally interesting. As the main sequence star accretes more mass, its internal pressure, and hence temperature, will have to increase so increasing its core hydrogen fusion rate. When, in the future, it exhausts the hydrogen in its core it too will expand into a red giant. As you might guess, at this point it will then begin to transfer mass back to the 0.8 solar mass star – a stellar see-saw!

## 7.8    High mass stars in the range >8 solar masses

Stars in this mass range have sufficient mass overlying the core so that the temperature of the core can increase beyond that in less massive stars. This allows the capture of alpha particles to proceed further. Having made carbon and oxygen, it is then possible to build up the heavier elements having atomic numbers increasing by four – produced by the absorption of alpha particles. Hence, the $^{16}$O fuses to $^{20}$Ne, the $^{20}$Ne fuses to $^{24}$Mg and then $^{24}$Mg fuses to $^{28}$Si producing a core dominated by silicon.

For each successive reaction to take place the temperature has to increase as there is a greater potential barrier for the incoming alpha particle to tunnel through. In the melee, protons can react with these elements to form nuclei of other atoms with intermediate atomic numbers, such as $^{19}$Fl and $^{23}$Na, though these will be less common. A shell like structure results, with layers of the star containing differing elements, the heaviest nearest to the centre.

When the temperatures reach the order of $3 \times 10^9$ K, silicon can be transformed though a series of reactions passing through $^{32}$S, $^{36}$A and continuing up to $^{56}$N. The silicon burning produces a core composed mostly of iron (the majority) and nickel. Iron and its close neighbours in the atomic table have the most stable nuclei, and any further reactions to build up heavier nuclei are endothermic (they would absorb energy rather than provide it) so this is where nuclear fusion has to stop. The star is then said to have an iron core. This core is surrounded by shells in which the lighter elements are still burning giving an interior like that shown in Figure 7.7.

**Figure 7.7** The 'onion-like' shells of fusion burning during the later stages in the evolution of a giant star. Image: Wikipedia Commons.

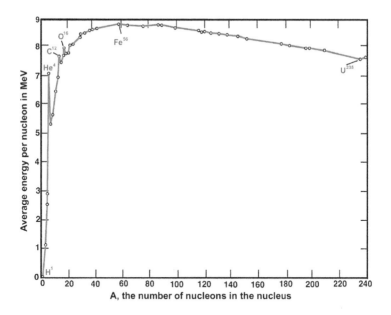

**Figure 7.8** The binding energy curve.

The energy released by each stage of burning is reduced and, as the element involved lies nearer to the peak of the binding energy curve (Figure 7.8), progressively less energy is released per gram of fuel. As a result, the time spent carrying out each successive reaction becomes shorter: a star of mass 20 times that of our Sun will spend about 10 million years on the main sequence burning hydrogen to helium, then spend about 1 million years burning helium to carbon and about 300 years burning carbon to oxygen. The oxygen burning takes around 200 days and silicon burning is completed in just 2 days!

Once the core reaches its iron state, things progress very rapidly. At the temperatures that exist in the core (of order $8 \times 10^9$ K for a 15 solar mass star) the photons have sufficient energy to break up the heavy nuclei, a process known as photodisintegration. An iron nucleus may produce 13 helium nuclei in the reaction:

$$^{56}\text{Fe} + \gamma \rightarrow 13 \, ^4\text{He} + 4\text{n}$$

These helium nuclei then break up to give protons and neutrons:

$$^4\text{He} + \gamma \rightarrow 2 \, \text{p}^+ + 2\text{n}$$

As energy is released when the heavy elements were produced, these inverse processes are highly endothermic (requiring energy to progress) and thus the temperature drops catastrophically. There is then not sufficient pressure to support the core of the star which begins to collapse to form what is called a neutron star.

In the forming neutron star, free electrons combine with the protons produced by the photodisintegration of helium to give neutrons, in the reaction:

$$p^+ + e^- \rightarrow n + \nu_e$$

The electron neutrinos barely interact with the stellar material, so can immediately leave the star carrying away vast amounts of energy – the neutrino luminosity of a 20 solar mass star exceeds it photon luminosity by seven orders of magnitude for a brief period of time! The outer parts of the core collapses at speeds up to $70\,000\,\mathrm{km\,s^{-1}}$ and, within about 1 s, the core, whose initial size was similar to the Earth, is compressed to a radius of about 40 km! This is so fast that the outer parts of the star, including the oxygen, carbon, and helium burning shells, are essentially left suspended in space and begin to infall towards the core.

The core collapse continues until the density of the inner core reaches about three times that of an atomic nucleus ($\sim 8 \times 10^{14}\,\mathrm{g\,cm^{-3}}$). At this density, the strong nuclear force, which in nuclei is attractive, becomes repulsive – an effect caused by the operation of the Pauli exclusion principle to neutrons and termed neutron degeneracy pressure. As a result of this pressure, the core rebounds and a shock wave is propagated outwards into the infalling outer core of the star. As the material above is now so dense, not all the neutrinos escape immediately and give the shock front further energy which then continues to work its way out to the surface of the star – there producing a peak luminosity of roughly $10^9$ times that of our Sun. This is comparable with the total luminosity of the galaxy in which the star resides!

## 7.9     Type II supernova

This sequence of events is called a Type II supernova. The peak absolute magnitude of about −18 then drops by around six to eight magnitudes per year so that it gradually fades from view. We believe that such supernovae will occur in our Galaxy on average about once every 44 years. Sadly, the dust in the plane of the galaxy only allows us to see about 10–20% of these and so they are not often seen.

### 7.9.1    *The Crab Nebula*

On July 4, 1054 AD a court astrologer during the Sung dynasty, Yang Wei-T'e, observed a supernova in the constellation Taurus. The gas shell thrown out in the supernova explosion was first discovered in modern times by John Bevis in 1731, who included it in his sky atlas, *Uranographia Britannica*. Later, in 1758, it was independently discovered by Charles Messier whilst he was searching for the return of Halley's Comet. It became the first object in the Messier catalogue with the name M1. The Third Earl of Rosse, who drew its form using his 72 in. telescope in Ireland, thought that it appeared similar to a horseshoe crab and so he called it the 'Crab Nebula', the name by which it is usually known (Figure 7.9).

The Crab Nebula is still, nearly 1000 years after it was first observed, expanding at a rate of $1500\,\mathrm{km\,s^{-1}}$ and its luminosity is about 10 000 times brighter than our Sun. Much of this radiation appears to be the result of electrons, moving close to the speed of light (called relativistic electrons), spiralling around magnetic field lines in the nebula. The fact that the nebula still appears so energetic remained a puzzle until a neutron star (which is the remnant of the stellar core) was discovered in 1969 at the centre of the nebula. This will be described in detail below. The gas shell, now of order $6 \times 4$ arcmin in size and shining at 8.4 magnitudes, can still be observed with a small telescope.

The Crab Nebula is thought to be the remains of a Type II supernova. A supernova, (1987A) that was observed in the nearby galaxy, the Large Magellanic Cloud, in 1987 is also thought to have been a Type II supernova, but those observed by Tycho Brahe in 1572 and Johannes Kepler in 1604 are thought to

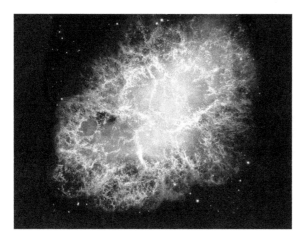

**Figure 7.9**  The Crab Nebula. Image: Hubble Space Telescope, NASA, ESA, J. Hester, A. Loll (ASU).

have been caused by a different mechanism and are termed Type I supernovae. These have proved very valuable for extending the cosmic distance scale out to distant galaxies and will be discussed in Chapter 9.

### 7.9.2 *Supernova 1987A*

In Febuary 1987, a supernova was observed in the Large Magellanic Cloud, a galaxy close to our own Milky Way (Figure 7.10). Visible for a while to the unaided eye, it became the closest observable supernova since that of 1604.

Supernova 1987A has played an important role in determining the cosmic distance scale as will be discussed in Chapter 8. However, there is a further aspect of its explosion that merits mention which was the result of a wonderful piece of serendipitous timing. In the late 1970s a particle physics model called the Grand Unified Theory (GUT) suggested that protons would decay with a half-life of $10^{31}$ years. This means that if one observed a number of protons for $10^{31}$ years half would have decayed. This is obviously not an experiment that can be mounted, but the possible proton decay could be detected if one observed a very large number of protons for a relatively short period. The proposed decay process is:

$$p^+ \rightarrow \varepsilon^+ + \pi^0$$
$$\pi^0 \rightarrow 2\gamma$$
$$\varepsilon^+ + \varepsilon^- \rightarrow 2\gamma$$

where $\varepsilon^+$ is a positron and $\pi^0$ is a neutral pi-meson or pion.

**Figure 7.10** Supernova 1987A in the Large Magellanic Cloud. Image: European Southern Observatory.

The proton decays into a positron and neutral pion which then immediately decays into two gamma rays. The positron will annihilate with an electron to form two more gamma rays.

To this end, a number of detectors were built in the 1980s including that at the Kamioka Underground Observatory located 1000 m below ground in Japan. To provide the protons, 3000 tons of pure water was contained in a cylinder 16 m tall and 15.6 m in diameter. The cylinder was surrounded by 1000 photomultiplier tubes attached to its inner surface which would be able to detect the gamma rays produced in the proton decay. It came into operation in 1983 and was given greater sensitivity in 1985. To date, even with a new detector containing 50 000 tons of water, no convincing proton decays have been detected and later versions of GUT suggest that the decay half-life might be nearer to $10^{35}$ years. However, what is critically important was that the detector, which came into full operation at the end of 1986 after its upgrade in 1985, could also detect neutrinos.

Relativity states that no particle can travel at the speed of light in a vacuum. However, in dense media, like water, light travels at lower speeds. It is thus possible for a particle to travel through water faster than the speed of light. If the particle is charged, it will emit light radiation called Cherenkov radiation. The process is analogous to the formation of a sonic boom when an airplane exceeds the speed of sound. Neutrino interactions with the electrons in the water can transfer almost all the neutrino momentum to an electron which then moves at relativistic speeds in the same direction.

The relativistic electron produces Cherenkov radiation which can be detected by the photomultiplier tubes around the tank. The expanding light cone will trigger a ring of photomultiplier tubes whose position gives an indication of the direction from which the neutrino has travelled. This makes it more than just a detector – it forms a very crude telescope!

When Supernova 1987A was seen to explode just a few months later (this being the serendipitous timing referred to earlier) the Kamiokande experiment detected 11 neutrinos within the space of 15 s. A similar facility in Ohio detected a further 8 neutrinos within just 6 s and a detector in Russia recorded a burst of 5 neutrinos within 5 s. These 24 neutrinos are the only ones ever to have been detected from a supernova explosion. Perhaps surprisingly, the neutrinos were detected some 3 h before the supernova was detected optically. This is not because they had travelled faster than light! They, of course, travelled out directly from the collapsing core of the star, whereas the visible light was not emitted until later when the shock wave reached the surface of the star. The detection of those 24 neutrinos was perfect confirmation of the theoretical models that had been developed for the core collapse of a massive star and consistent with theoretical prediction that ~$10^{58}$ neutrinos would be produced in such an event. The Kamiokande observations also allowed an upper limit to be placed on the neutrino mass. If one assumes that the neutrinos began their trip somewhat ahead of the light from the supernova and given the fact that they arrived before the light having travelled through space for ~169 000 years means that they must have been travelling very close (within one part in $10^8$) to the speed of light. This, together with the fact that the higher and lower energy neutrinos arrived at the same time allows an upper limit to be put on the mass of a neutrino. It cannot be greater than 16 eV which is about three-millionths the mass of an electron. (Other evaluations of the data give a somewhat lower value of ~7 eV.)

**Electronvolt**

The unit, called an electronvolt (eV), is often used as an indicator of mass when talking about objects like protons, neutrons and electrons. An electronvolt is a measurement of energy, but since mass and energy are related by Einstein's famous equation, $E = mc^2$, the energy equivalent of mass may be used instead. In terms of MeV (1 MeV = 1 000 000 eV), the masses are:

$$\text{Neutron} = 939.56563 \text{ MeV}$$
$$\text{Proton} = 938.27231 \text{ MeV}$$
$$\text{Electron} = 0.51099906 \text{ MeV}$$

## 7.10    Neutron stars and black holes

What remains from this cataclysmic stellar explosion depends on the mass of the collapsing core. When stars, whose total mass is greater than ~8 solar masses but less than ~12 solar masses, collapse the result is a neutron star – the core being supported by neutron degeneracy pressure as described above. The typical mass of such a neutron star would be ~1.4 solar masses so that it is, in effect, a giant nucleus containing ~$10^{57}$ neutrons. It will have a radius between 10 km and 15 km – the theoretical models are not all that precise. Assuming a radius of 10 km, the average density would be $6.65 \times 10^{14}$ g cm$^{-3}$ – more than that of an atomic nucleus!

Gravity at the surface would be intense; for a 1.4 solar mass star with a radius of 10 km, the acceleration due to gravity at the surface would be 190 billion times that on the surface of the Earth and the speed of an object having fallen from a height of 1 m onto the surface would be 6.88 million km h$^{-1}$! A simple Newtonian calculation of the escape velocity from the surface gives a value of $0.643c$. This implies that both special and general relativity need to be invoked when considering neutron stars. The structure of a neutron star is very complex; part may even be in the form of a superfluid sea of neutrons which will thus have no viscosity. This can give rise to an observable consequence as will be described later.

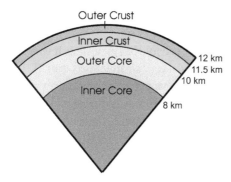

**Figure 7.11** Cross-section of a neutron star. Outer (solid) crust: nuclei of iron and nickel; inner crust: nuclei, superfluid electrons and electrons; outer core: superfluid neutrons, superconducting protons and electrons; inner core: condensed pions, kaons and quark matter?

A neutron star may have an outer crust of heavy nuclei, the majority being of iron and nickel. Within this is an inner crust containing elements such as krypton, superfluid neutrons and relativistic degenerate electrons. The inner crust overlays an interior of superfluid neutrons intermixed with superconducting protons and relativistic degenerate electrons. Finally there may be a core of pions or other elementary particles (Figure 7.11).

Like white dwarfs, neutron stars become smaller and denser with increasing mass, but there will become a point when the neutron degeneracy pressure can no longer support the mass of the star. So, in an analogous manner to the Chandrasekhar Limit for the maximum mass of a white dwarf, there is a limit, believed to be about 3 solar masses, beyond which the collapse continues to form a black hole as will be discussed in Section 7.13.

Stars rotate as, for example, our Sun which rotates once every ~25 days at its equator. The core of a star will thus have angular momentum. As the core collapses, much of this must be conserved (some is transferred to the surrounding material), so the neutron star that results will be spinning rapidly with rotational periods of perhaps a few milliseconds. The neutron star will also be expected to have a very intense magnetic field. This rotating field has observational consequences that have allowed us to discover neutron stars and investigate their properties.

When the neutron star is first born its surface temperature may approach $10^{11}$ K. It initially cools by emitting neutrinos and antineutrinos – an interesting process that lasts about a day. A neutron decays to a proton, electron and an antineutrino. The proton then combines with an electron to give a neutron and a neutrino – a sort of merry-go-round during which the neutrinos carry away energy and cool the star down to about $10^9$ K. Neutrinos carry away much of the star's energy for about 1000 years whilst the surface temperature falls to a few million kelvin. Photons – in the form of X-rays – then carry energy away from the surface which stays close to 1 million K for the next thousand years. Its luminosity will then be comparable with that of our Sun.

This explains why the Crab Nebula could be still visible. It was known that a star close to the centre of the nebula had a very strange spectrum. If this were the neutron star associated with the supernova explosion, its energy output would have kept the gas thrown out into the interstellar medium excited, so remaining visible. The way in which this was confirmed and how, to date, nearly 2000 neutron stars have been discovered is one of the most interesting stories of modern astronomy.

## 7.11    The discovery of pulsars

When observations of stars through the Earth's atmosphere were described earlier it was pointed out that stars scintillate ('twinkle' is a rather nice if not scientific term that is often used). This is because irregularities in the atmosphere passing between the observer and the star act like alternate convex and concave lenses which sequentially converge the light from the star (so making it appear slightly brighter) and then diverge it (so reducing its brightness). It was pointed out that planets do not scintillate to any significant degree. Their angular size means that the light is passing through a large number of adjacent atmospheric cells and the effects average out.

There is a similar effect related to radio sources caused by irregularities in the solar wind – bubbles of gas which stream out from the Sun expanding as they do so. It was realised that this could lead to a way of investigating the angular sizes of radio sources by studying the amount of scintillation observed when the source was at different angular distances from the Sun. It would also be a way of discovering radio sources with very small angular sizes such as quasars which will be described in Chapter 8. To carry out this experiment a very large antenna was required and Tony Hewish at the Mullard Radio Astronomy Laboratories at Cambridge recruited a PhD student called Jocelyn Bell to first help build the antenna (made up of an array of 2048 dipoles) and then carry out and analyse the observations (Figure 7.12). The array observed radio sources as they passed due south so a given radio source would be observed every sidereal day as it appeared on the meridian.

The signals from radio sources appeared on a roll of chart, about 400 ft (~122 m) of which was produced each day. Soon Bell was able to distinguish between the scintillating signal of a radio source and interference, often from cars passing the observatory. In July she observed a 'little bit of scruff' that did not look like a scintillating radio source but did not appear like interference either. A second intriguing feature was that it had been observed at night when a radio source would be seen away from the direction of the Sun and scintillation would not be expected to be seen. Looking through the charts, she discovered that a similar

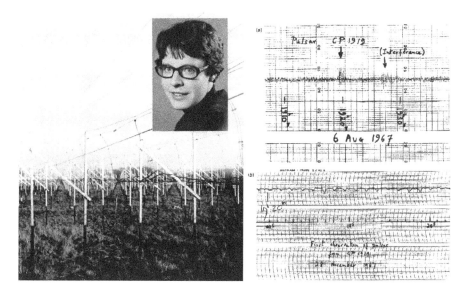

**Figure 7.12** Jocelyn Bell with the Cambridge array (which discovered the first pulsar) and the discovery record. Image: Jocelyn Bell and Tony Hewish, University of Cambridge.

signal had been seen earlier from the same location in the sky. She observed that it reappeared again at always a precise number of sidereal days later which implied that the radio source, whatever it was, was amongst the stars rather that within the Solar System. Hewish and Bell then equipped the receiver with a high speed chart recorder to observe the 'scruff' in more detail and discovered to their amazement that it was not random, but a series of precisely spaced radio pulses having a period of $1.33724$ s.

Observations using a different telescope at Cambridge confirmed the presence of the signal and also the fact that the pulse arrived at slightly different times as the frequency of observation was changed. This effect is called dispersion, and is exactly similar to the fact that different wavelengths of light travel at different speeds in glass. The interstellar medium is not a perfect vacuum and so can cause this effect but it would be only observed if the source of the pulses was far beyond the Solar System.

At that time, no one in the radio astronomy group at Cambridge group could conceive of a natural phenomenon that could give rise to such highly precise periodic signals – it seemed that no star, not even a white dwarf could pulsate at such a fast rate – and they wondered if it might be a signal from an extraterrestrial civilization. Bell, who called the source LGM1 (Little Green Men 1), was somewhat annoyed about this as it was disrupting her real observations. When, later, a second source with similar characteristics but a slight faster period of $1.2$ s was discovered she was somewhat relieved as 'it was highly unlikely that two lots of Little Green Men could choose the same unusual frequency and unlikely technique to send a signal to the same inconspicuous planet Earth!'

A few days before the paper presenting these discoveries was published in *Nature* in February 1968, Hewish announced the discovery to a group of astronomers at Cambridge. Fred Hoyle was amongst them, and suggested that the signal might be pulsed emissions coming from an oscillating neutron star – the theoretical remnant of a supernova but never previously observed. After a press conference following the publication of the *Nature* paper announcing the discovery, the science correspondent of the *Daily Telegraph* coined the name pulsar for these enigmatic objects.

Some 3 months later, in a paper also published in *Nature*, Thomas Gold at Cornell University in Ithaca, USA, gave an explanation for the pulsed signals. Gold suggested that the radio signals were indeed coming from neutron stars, but that the neutron star was not oscillating, but instead spinning rapidly around its axis. He surmised that the rotation, coupled with the expected intense magnetic field generates two steady beams of radio waves along the axis of the magnetic field lines, one beam above the north magnetic pole and one above the south magnetic pole. If (as in the case of the Earth) the magnetic field axis is not aligned with the neutron star's rotation axis, these two beans would sweep around the sky rather like the beam from a lighthouse (Figure 7.13). If then, by chance, one of the two beams crossed our location in space, our radio telescopes would detect a sequence of regular pulses – just as Bell had observed – whose period was simply the rotation rate of the neutron star.

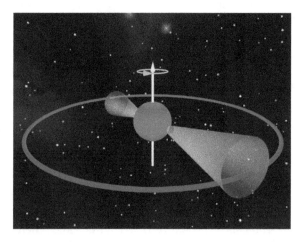

**Figure 7.13**  Twin beams emitted by a pulsar. Image: Michael Kramer, University of Manchester.

Gold, in this paper, pointed out that a neutron star (due to the conservation of angular momentum when it was formed) could easily be spinning at such rates. He expected that most pulsars should be spinning even faster than the first two observed by Jocelyn Bell and suggested a maximum rate of around 100 pulses per second.

Since then, nearly 2000 pulsars have been discovered. The majority have periods between 0.25 s and 2 s. It is thought that as the pulsar rotation rate slows the emission mechanism breaks down and the slowest pulsar detected has a period of 4.308 s. There is a class of 'millisecond' pulsars where the proximity of a companion star has enabled the neutron star to 'pull' material from the outer envelope of the adjacent star onto itself. This also transfers angular momentum so spinning the pulsar up to give periods in the millisecond range – hence their name. The fastest known pulsar is spinning at just over 700 times per second – with a point on its equator moving at 20% of the speed of light and close to the point where it is thought theoretically that the neutron star would break up!

Pulsars slowly radiate energy, which is derived from their angular momentum. This is so high that the rate of slowdown is exceptionally slow and consequently pulsars make highly accurate clocks and some may even be able to challenge the accuracy of the best atomic clocks. The periods of all pulsars slowly increase (except when being spun up to form a millisecond pulsar) and a typical pulsar would have a lifetime of a few tens of millions of years.

The linking of pulsars with supernova neutron star remnants was confirmed when the 'odd' star close to the centre of the Crab Nebula was shown to be a pulsar with a period of 0.0333 s – rotating just over 30 times per second. A second pulsar was discovered within the Vela supernova remnant and both this and the Crab Pulsar also emit beams of radiation not just at radio waves, but across the whole electromagnetic spectrum including visible light, X-rays and gamma rays.

### 7.11.1     *What can pulsars tell us about the universe?*

Pulsars give us a way of investigating matter in a super-dense form that we have no possibility of making here on Earth. For example, observations of the Crab Pulsar have enabled verification of aspects of the theoretical makeup of neutron stars described above. A small radio telescope at the Jodrell Bank Observatory observes the pulses from the Crab Pulsar all the time that it lies above the horizon. It has been discovered that, about every 3–6 years, its spin rate suddenly increases – an event known as a glitch. This seems odd as generally pulsars slow down with time. The likely explanation is this: as the outer parts of the neutron star slow, an inner part, thought to be in the form of a sea of superfluid neutrons (and thus having no viscosity) would not be slowed as it is effectively decoupled from the outer parts of the neutron star. The situation will then arise that the superfluid region towards the centre of the neutron star will be rotating more rapidly than the outer parts. It is thought this region can contain vortices and eventually, when these interact with the outer parts of the star, angular momentum can be transferred which rapidly increases the pulsar's rotation rate! The pulsar then continues to slow with the centre remaining at a constant (but now slower) rotation rate until the process repeats.

As the pulse travels through the interstellar medium, the radio waves cause electrons to vibrate in sympathy. This process, called dispersion, slows the radio waves and has a greater effect at lower frequencies. A sharp pulse emitted by the pulsar will thus be gradually broadened as it travels through space and, if observed over a range of frequencies, the pulse is seen to arrive earlier at higher frequencies as shown in Figure 7. 14. The amount of dispersion depends on the total electron content along the route the pulse has taken to the Earth and so more distant pulsars will show more dispersion. If one assumes the electron density in the interstellar medium is roughly constant, one can use the measured dispersion to estimate the distances to the pulsars.

Most pulsars are seen along the plane of the galaxy, just as one would suspect as they are the remnants of stars but, perhaps surprisingly, a significant number are observed away from the plane. The 217-km MERLIN array at Jodrell Bank Observatory is capable of making very precise measurements of the position of pulsars and has observed, from positional measurements made over a number of years, that many are moving at speed comparable with, and even exceeding, the

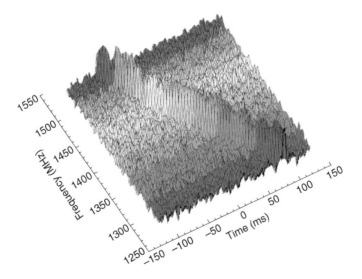

**Figure 7.14** Diagram showing the dispersion of a radio pulse. Image: University of Manchester.

escape velocity of the galaxy ($\sim 500\,\mathrm{km\,s^{-1}}$). The highest pulsar speed so far measured (in this case by the USA's 5000-km VLBA array) is $1100\,\mathrm{km\,s^{-1}}$ – London to New York in 5 s!

These pulsars have obviously been ejected from the supernova explosion that gave rise to them with very great energies, enabling them to travel around the galaxy and, in some cases, to leave the galaxy into the depths of intergalactic space. It appears that, usually, the supernova explosion will be more intense on one side or the other of the central neutron star which is then ejected at high speeds rather like a bullet from a gun. In some cases it is even possible to track the course of a pulsar back to the gaseous remnant of the supernova. The situation where the resulting pulsar remains within the supernova gas shell, such as in the Crab Nebula, appears to be very rare.

## 7.12    Pulsars as tests for general relativity

Pulsar systems have played a very significant role in testing Einstein's General Theory of Relativity – his theory of gravity. It is the fact that pulsars are such superbly accurate clocks that have made them such valuable tools with which to test Einstein's theory. The first such test came with the discovery, by Russell Taylor and Joseph Hulse in 1974, of the first 'binary pulsar'. In the binary pulsar system a 1.4 solar mass pulsar is orbiting a companion star of equal mass. It thus comprises two co-rotating stellar mass objects. General relativity predicts that such a system will radiate gravitational waves – ripples in space–time that propagate out through the universe at the speed of light. Though gravitational wave detectors are now in operation across the globe, this gravitational radiation is far too weak to be directly detected. However, there is a consequent effect that *can* be detected. As the binary system is losing energy as the result of its gravitational radiation, the two stars should gradually spiral in towards each other. The fact that one of these objects is a pulsar allows us to very accurately determine the orbital parameters of the system. Precise observations over the 40 years since it was first discovered, shown in Figure 7.15, show how the two bodies are slowly spiralling in towards each other, exactly agreeing with Einstein's predictions! Taylor and Hulse received the Nobel Prize for Physics in 1993 for this outstanding work.

It is another pulsar system, this time where both objects in the system are pulsars, and called the 'double pulsar', that has produced one of the most stringent tests of general relativity to date (Figure 7.16). It was discovered in a survey carried out at the Parkes Telescope in Australia using receivers and data acquisition equipment built at the University of Manchester's Jodrell Bank Observatory. In analysis of the resulting data using a super-computer at Jodrell Bank the double pulsar

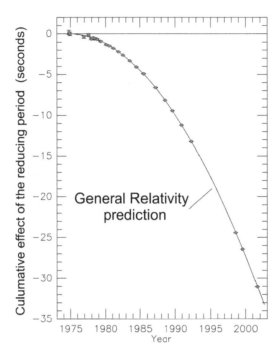

**Figure 7.15** The cumulative effect on the period of the orbit on the objects in the binary pulsar system due to the emission of gravitational waves. (The break in observations following 1992 was due to a major upgrade to the Arecibo Telescope.)

**Figure 7.16** The double pulsar system, showing the distortion of space–time resulting from their mass. Image: Michael Kramer, University of Manchester.

was discovered in 2003. It comprises two pulsars of 1.25 and 1.34 solar masses spinning with rotation rates of 2.8 s and 23 ms, respectively. They orbit each other every 2.4 h with an orbital major axis just less than the diameter of the Sun.

The neutron stars are moving at speeds of 0.01% that of light and it is thus a system in which the effects of general relativity are more apparent than in any other known system. At this moment in time, general relativity predicts that the two neutron stars should be spiralling in towards each other at a rate of 7 mm per day. Observations made across the world since then have shown this to be exactly as predicted. In fact, five predictions of general relativity can be tested in this unique system. The one that has so far provided the highest precision is a measurement of what is called the Shapiro delay, which was discussed in Section 2.9.2.

By great good fortune, the orbital plane of the two pulsars is almost edge on to us. Thus, when one of the two pulsars is furthest away from us its pulses have to pass close to the nearer one on their way to our radio telescopes. They will thus have to travel a longer path through the curved space surrounding the nearer one and suffer a delay that is close to 92 μs. The timing measurements agree with theory to an accuracy of 0.05%. Einstein must be at least 99.95% right!

## 7.13   Black holes

If one projected a ball vertically from the equator of the Earth with increasing speed, there comes a point, when the speed reaches 11.2 km s$^{-1}$, when the ball would not fall back to Earth but would escape the Earth's gravitational pull. This is the Earth's escape velocity. If either the density of the Earth was greater or its radius smaller (or both) then the escape velocity would increase as Newton's formula for escape velocity shows:

$$v_0 = \sqrt{\frac{2GM}{r_0}}$$

where $v_0$ is the escape velocity, $M$ the mass of the object, $r_0$ its radius and $G$ the universal constant of gravitation.

If one naively used this formula in realms where relativistic formula would be needed, one could predict the mass and/or size of an object where the escape velocity would exceed the speed of light and thus nothing, not even light, could escape. The object would then be what is termed a black hole.

Black holes have no specifically defined size or mass, but so far we have only found evidence for black holes in two circumstances. The first, with masses of up to a billion or more times that of our Sun, are found the heart of galaxies and will be discussed in Chapter 8. The second are believed to result from the collapse of a stellar core whose mass exceeds ~3 solar masses – the point at which neutron degeneracy pressure can no longer prevent gravitational collapse.

The surface surrounding the remnant within which nothing can escape is called the event horizon. In the simplest case when the black hole is not rotating, the event horizon is the surface of a sphere and has a radius, called the Schwarzschild radius, given by:

$$R_S = 2GM/c^2$$

The interior of an event horizon is forever hidden from us, but Einstein's theories predict that at the centre of a non-rotating black hole is a singularity, a point of zero volume and infinite density where all of the black hole's mass is located and where space–time is infinitely curved. This author does not like singularities; in his view they are where the laws of physics are inadequate to describe what is actually the case. We know that somehow, Einstein's classical theory of gravity must be combined with quantum theory and so, almost certainly, relativity *cannot* predict what happens at the heart a black hole.

As will be covered in more detail in the description of the Big Bang in Chapter 9, nucleons are thought to be composed of up quarks and down quarks. It is possible that at densities greater than those that can be supported by neutron degeneracy pressure, quark matter could occur – a degenerate gas of quarks. Quark-degenerate matter *may* occur in the cores of neutron stars and *may* also occur in hypothetical quark stars. Whether quark-degenerate matter can exist in these situations depends on the, poorly known, equations of state of both neutron-degenerate matter and quark-degenerate matter.

Some theoreticians even believe that quarks might themselves be composed of more fundamental particles called preons and if so, preon-degenerate matter might occur at densities greater than that which can be supported by quark-degenerate matter. Could it be that the matter at the heart of a black hole is of one of these forms?

The more massive a black hole, the greater the size of the Schwarzschild radius: a black hole with a mass 10 times greater than another will have a radius 10 times as large. A black hole of 1 solar mass would have a radius of 3 km, so a typical 10 solar mass stellar black hole would have a radius of 30 km.

### 7.13.1    *The detection of stellar mass black holes*

If a stellar black hole, formed when a massive star ends its life in a supernova explosion, existed in isolation, it would be very difficult to detect: gravitational microlensing, a method now being employed to detect planets, might just be

**Figure 7.17**  Material accreting on to a black hole from a companion star.
Image: Thierry Lombry.

able to do so. However, many stars exist in binary systems. In a binary system in which one of the components is a black hole, it appears that its gravity can pull matter off the companion star forming an accretion disc of gas swirling into the black hole (Figure 7.17). As gas spins up as it nears the black hole due to conservation of angular momentum, the differential rotation speeds give rise to friction and the matter in the accretion disc reaches temperatures of more than 1 million K. It thus emits radiation, mostly in the X-ray part of the spectrum.

X-ray telescopes have now detected many such X-ray binary systems, some of which are thought to contain a black hole. Observations of the orbital size and velocity of the normal star in the system enable one to estimate the mass of its companion. If this is both invisible and exceeds a calculated mass of ~3 solar masses, then it is likely to be a black hole. An excellent candidate in our own Galaxy is Cygnus X-1 – so-called because it was the first X-ray source to be discovered in the constellation Cygnus and is the brightest persistent source of high energy X-rays in the sky. Usually called Cyg X-1, it is a binary star system that contains a super-giant star with a surface temperature of 31 000 K (with its spectral type lying on the O and B boundary) together with a compact object. The mass of the super-giant is from 20 to 40 solar masses and observations of its orbital parameters imply a companion of 8.7 solar masses. This is well above the 3 solar mass limit of a neutron star, so it is thought to be a black hole.

### 7.13.2    *Black holes are not entirely black*

In the 1970s, Stephen Hawking showed that due to quantum-mechanical effects, black holes actually emit radiation – they are not entirely black! The energy that produces the radiation in the way described below comes from the mass of the black hole. Consequently, the black hole gradually loses mass and, perhaps surprisingly, the rate of radiation increases as the mass decreases, so the black hole continues to radiate with increasing intensity losing mass as it does so until it finally evaporates.

The theory describing why this happens is highly complex and results from the quantum mechanical concept of virtual particles – mass and energy can arise spontaneously provided they disappear again very quickly and so do not violate the Heisenberg Uncertainty Principle. In what are called vacuum fluctuations, a particle and an antiparticle can appear out of nowhere, exist for a very short time, and then annihilate each other. Should this happen very close to the event of a black hole, it can sometimes happen that one particle falls across the horizon, while the other escapes. The particle that escapes carries energy away from the black hole and can, in principle, be detected so that it appears as if the black hole was emitting particles.

Black holes can be said to have an effective temperature, and unless this is less than the temperature of the universe the black hole cannot evaporate. This temperature is now ~2.7 K – the remnant of the radiation left over from the Big Bang which will be discussed in Chapter 9 – and is vastly higher than the effective temperatures of even solar mass black holes. Eventually, in aeons, when the temperature of this relict radiation has fallen sufficiently and assuming Hawking's theory is correct, stellar mass black holes may finally begin to evaporate – on a timescale of $10^{100}$ years!

## 7.14    Questions

1.  A circular planetary nebula (like the Ring Nebula in Figure 7.5) has an angular size of 2 arcmin and lies at a distance of ~600 pc. Doppler shift measurements show that the gaseous shell is expanding at a rate of 20 km s$^{-1}$. How long ago was the planetary nebula formed?

2.  The Ring Nebula has an angular diameter of 72 arcsec. Given that it is at a distance of ~1500 pc, estimate its diameter. It is expanding at a velocity of 15 km s$^{-1}$. Estimate its age.

3.  Betelgeuse is a red giant at a distance of 428 light years. In the future it will become supernova similar to 'Tycho's supernova' which was observed in 1572 and lies at a distance of 9800 light years. At its peak, its brightness

was similar to that of Venus (which has a peak magnitude of $\sim-4$). What might we expect the peak apparent magnitude of the Betelgeuse supernova explosion to be?

4.    A neutron star has a radius of $11.5\,km$ and a mass of about $1.35M_{sun}$. Calculate the density of this neutron star in kilograms per cubic centimetre. Compare the mass of Mount Everest ($\sim5 \times 10^{10}\,kg$) with the mass with $1\,cm^3$ of neutron star material. ($M_{sun} = 2 \times 10^{30}\,kg$.)

5.    A neutron star with a radius of $10\,km$ is observed to rotate $712$ times per second. At what speed, relative to that of light, is the surface of the neutron star's equator moving?

6.    The Schwarzschild radius, $R_{S.}$ of a black hole is given by $2GM/c^2$. Calculate $R_S$ for a stellar remnant of $10$ solar masses which has become a black hole.

## Chapter 8

# Galaxies and the Large Scale Structure of the Universe

## 8.1 The Milky Way

On a dark night with transparent skies, we can see a band of light across the sky that we call the Milky Way. (This comes from the Latin *Via Lactea*.) The light comes from the myriads of stars packed so closely together that our eyes fail to resolve them into individual points of light. This is our view of our own Galaxy, called the Milky Way Galaxy or often 'the Galaxy' for short. It shows considerable structure due to obscuration by intervening dust clouds. The band of light is not uniform; the brightness and extent is greatest towards the constellation Sagittarius suggesting that in that direction we are looking towards the galactic centre. However, due to the dust, we are only able to see about one-tenth of the way towards it. In the opposite direction in the sky the Milky Way is less apparent implying that we live out towards one side. Finally, the fact that we see a band of light tells us that the stars, gas and dust that make up the Galaxy are in the form of a flat disc.

Figure 8.1 is the best all-sky image of our Galaxy ever taken, and the author is indebted to Axel Mellinger for allowing it to be used in this book. It is a composite from images taken from dark sky sites around the world and covers the night sky in the same way that a map can show the whole of the Earth's globe. Across the image is the sweep of the Milky Way, with the centre of the Galaxy, lying in the constellation Sagittarius, showing as a bulge in the exact centre of the image. Below and to the right of the centre are seen two nearby galaxies, the Large and Small Magellanic Clouds. The constellation Orion is at the right just below the path of the Milky Way. The Andromeda Galaxy can just be seen as a thin streak below the Milky Way two-thirds of the way to the left of the image from the centre.

The major visible constituent of the Galaxy, about 96%, is made up of stars, with the remaining 4% split between gas (~3%) and dust (~1%) (Figure 8.2). Here 'visible' means that detection can be done by electromagnetic radiation; visible, infrared or radio. As we shall see later, we suspect that there is a further component of the Galaxy that we cannot detect (called 'dark matter').

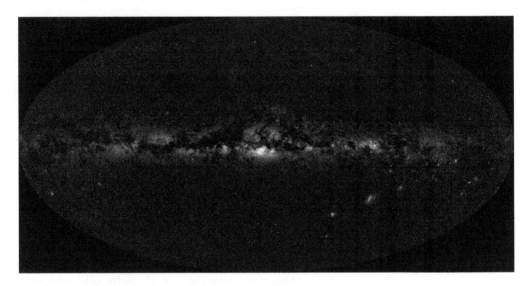

**Figure 8.1** All-sky image of the Milky Way. Image: Axel Mellinger.

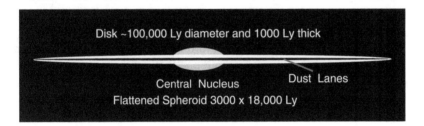

**Figure 8.2** Cross-section of our Galaxy.

### 8.1.1    *Open star clusters*

Amongst the general star background we can see close groupings of stars that are called clusters. These are of two types; open clusters and globular clusters. Open clusters are a consequence of the formation of a group of stars in a giant cloud of dust and gas and are thus naturally found along the plane of the Milky Way – the disc of our Galaxy. Over time the stars will tend to drift apart but, whilst they are young, we will see the stars relatively closely packed together. Prime examples observable in the northern hemisphere are the Hyades and Pleiades clusters in Taurus and the Double cluster in Perseus. (The Pleiades cluster is shown in Figure 2.1.)

One can get an indication of the age of a cluster by plotting a Hertzsprung–Russell diagram (H–R diagram) of the stars in the cluster. Figure 8.3 shows the

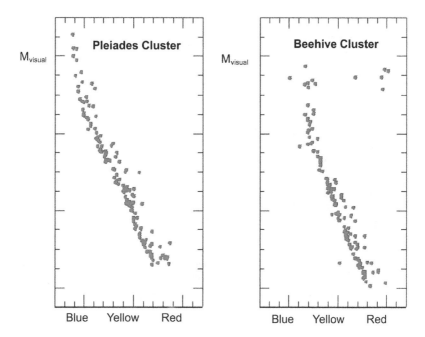

**Figure 8.3** Hertzsprung–Russell diagrams of two open clusters.

H–R diagrams of the Pleiades cluster, M45, and the Beehive cluster (or Praesepe, Latin for 'manger'), M44.

All stars in the Pleiades cluster lie on the main sequence – even the brightest, Alcyone, which is 1000 times more luminous than our Sun. As we have seen, such stars have relatively short lifetimes on the main sequence, and stellar evolution studies indicate that the Pleiades cluster cannot be more than ~115 million years old.

In contrast, the H–R diagram for the Beehive cluster shows that the stars that would have been at the upper left of the main sequence have evolved off it; five of which are now red giants in the upper right of the diagram and some, not shown, have reached the end of their lives as white dwarfs. Its age is believed to be ~750 million years.

## 8.1.2    *Globular clusters*

The globular clusters are, in contrast, very old stars in tight spherical concentrations (~200 light years across) of 20 000–1 million stars. In the northern hemisphere the most spectacular is M13 in Hercules (Figure 8.4), whilst Omega Centauri is a jewel of the southern hemisphere. They date from the origin of the Galaxy and were formed in the initial star formation period of our Galaxy but

**Figure 8.4** M13, the globular cluster in Hercules. Image: Robert J. Vanderbei, Wikipedia Commons.

their precise origin and role in the evolution of the Galaxy is still unclear. Globular clusters orbit the centre of our Galaxy and form a roughly spherical distribution helping to form what is known as the galactic halo. We know of 150 globular clusters associated with our Galaxy and perhaps a further 20 may be present but obscured by dust. Their spherical shape is due to the fact that they are very tightly bound by gravity, a further consequence of which is that the stars near their centres are very tightly packed.

### 8.1.3    *The interstellar medium and emission nebulae*

Together the gas and dust make up what is called the interstellar medium (ISM). Most of the ISM is not apparent to our eyes but in some regions we can see either 'emission nebulae' where the gas glows or 'dark nebulae' where a dust cloud appears in silhouette against a bright region of the galaxy. Perhaps the most spectacular example of an emission nebula is the Great Nebula in Orion, or more simply the Orion Nebula – a region of star formation where the hydrogen gas is being excited by the ultraviolet light emitted by the very hot young stars (forming the 'Trapezium') at its heart (Figure 8.5). This type of emission nebula is called an HII ('H-two') region as it contains ionized hydrogen where the electrons have been split off from the protons by the ultraviolet photons emitted by very hot stars.

**Figure 8.5** The Orion Nebula, M42. Image: NASA, ESA, M. Robberto (STScI/ESA) and The Hubble Space Telescope Orion Treasury Project Team.

The protons and electrons can then recombine to form neutral hydrogen atoms (HI or 'H-one') and the electrons drop through the allowed energy levels to the lowest energy state emitting photons of various wavelengths as they do so. One of these transitions gives rise to a bright red emission line at 6563 Å, so in photographs these regions look a pinky-red colour.

An example of a dark nebula is the 'Coal Sack', seen against the background of the Milky Way close to the Southern Cross. Often the two types are seen together, as in the Eagle Nebula in Serpens and the Horsehead Nebula in Orion, where the dark pillars of dust and the 'horse's head', respectively, are seen against the bright glow of excited gas clouds (Figure 8.6).

## 8.1.4 *Size, shape and structure of the Milky Way*

The size of the galaxy was first measured by Harlow Shapley who measured the distances to 100 of the globular clusters associated with our Galaxy. He found that they formed a spherical distribution, whose centre should logically be the

**Figure 8.6** The Eagle Nebula (a) and Horsehead Nebula (b). Images: NASA, Jeff Hester, and Paul Scowen (Arizona State University) (a) and Nigel Sharp (NOAO), KPNO, AURA, NSF (b).

centre of the galaxy, and so deduced that our Sun was ~30 000 light years distant from it and the diameter was about 100 000 light years across. Figure 8.7 shows a cross-section of the galaxy with the positions of the globular clusters observed by Shapley.

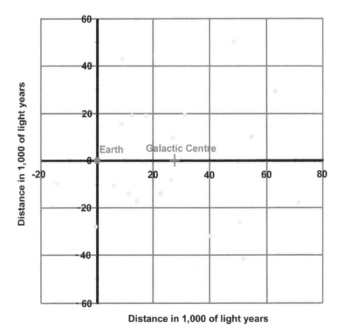

**Figure 8.7** The distribution of globular clusters as observed by Harlow Shapley.

We now believe that the Sun is 27.7 thousand years from the galactic centre and, using spectroscopic measurements to observe its motion relative to the
globular clusters, we can calculate that the Sun is moving around the centre of
the galaxy at about $220\,\mathrm{km\,s^{-1}}$, taking ~230 million years to travel once around
it. It appears that the central parts of the galaxy rotate like a solid body so that the
rotation speed increases as one moves out from the centre. Measurements of the
speed at which stars and gas rotate around the centre of the galaxy as a function
of distance produce what is called the 'galactic rotation curve'.

But what of its structure? Neutral hydrogen (HI) emits a radio spectral line
with a wavelength of 21 cm. Radio observations of this line along the plane of
the Milky Way show that the gas in the disc is not uniformly dense but is concentrated into clouds whose velocity away from or towards us can be determined
using the Doppler shift in its observed wavelength. These data can be used to plot
out the positions of the gas clouds and when this is done a pattern of spiral arms
emerges – indicating that we live in a typical spiral galaxy thought to be quite
similar to the nearby Andromeda Galaxy.

### 8.1.5    *Observations of the hydrogen line*

Let us consider this in more detail. A neutral hydrogen atom consists of a proton
and an electron. As well as their motion in orbit around each other, the proton and
electron also have a property called spin. This is actually a quantum mechanical
concept, but is analogous to the spin of the Earth and the Sun about their rotation
axes. Again, using this analogy, the spin may be clockwise or anticlockwise. Thus
their two spins may be oriented in either the same direction or in opposite directions. In a magnetic field (as exists in the Galaxy) the state in which the spins of
the electron and proton are aligned in the same direction has slightly more energy
than one where the spins of the electron and proton are in opposite directions.
Very rarely (with a probability of $2.9 \times 10^{-15}\,\mathrm{s^{-1}}$) a single isolated atom of neutral
hydrogen will undergo a transition between the two states and emit a radio spec-
tral line at a frequency of 1420.40575 MHz. This frequency has a wavelength of
~21 cm so that it is often called the 21-cm line. A single hydrogen atom will only
make such a transition in a timescale of order 10 million years but, as there are a
very large number of neutral hydrogen atoms in the ISM, it can be easily observed
by radio telescopes. This radio spectral line was first detected by Professor Edward
Purcell and his graduate student, Harold Ewen, at Harvard University in 1951
using a simple horn antenna.

Let us consider what we would see if our Galaxy had a number of spiral arms.
As well as delineating concentrations of stars and dust, they will also contain a
higher concentration of hydrogen gas. As all the material in the galaxy is rotating around its centre, when we look in different directions along the galactic

plane the material in the spiral arms – in some directions we might see several – will have differing velocities away from or towards us. The hydrogen line from these different arms will thus be Doppler shifted to higher or lower frequencies. Due to motion of the gas within an element of an arm that we observe, any given arm will give a line profile that is approximately Gaussian as shown in Figure 8.8a. This shows the gas close to us in our local spiral arm so its average velocity relative to us is zero. If the beam of our radio telescope receives radiation from the hydrogen gas in a spiral arm coming towards us we will see that Gaussian profile shifted to higher frequencies (Figure 8.8b), whilst if moving away we will observe it at lower frequencies (Figure 8.8c). In general, should we look in some arbitrary direction we may observe several arms and so get a more complex profile (Figure 8.8d) which can be 'dissected' into the individual profiles from each arm (Figure 8.8e).

Figure 8.9 shows some hydrogen line profiles observed using a 6.4-m radio telescope at the Jodrell Bank Observatory. Each has the galactic latitude at which it was observed; latitude 0 would mean that our telescope would be pointing

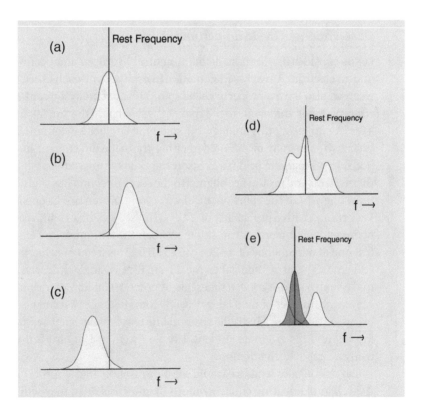

**Figure 8.8** Hydrogen line profiles.

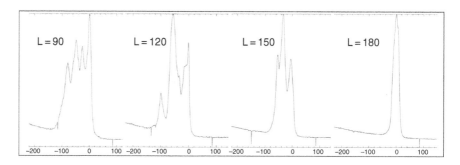

**Figure 8.9** Galactic hydrogen line profiles. Image: Christine Jordan, University of Manchester.

directly at the centre of the galaxy, latitude 180 directly away from it – looking out from our Galaxy. You can see that in this latter direction, there is only one peak in the spectrum which is centred on zero velocity with respect to us. This is the hydrogen in our local spiral arm which is all that we observe in this direction and is, of course, at rest with us. As we lie within our spiral arm there will always be a peak at zero relative velocity but, as we look along the Milky Way in other direc-tions, we see other peaks in the spectrum, corresponding to other spiral arms. For example, at latitude 90 (L = 90) we see three other spiral arms.

By observing such hydrogen line profiles along the galactic plane, and using a model of the rotation curve of the galaxy, it is possible to locate the position of each observed arm along the line of sight through the galaxy and build up a pic-ture of the spiral structure of our own Galaxy. Figure 8.10 shows a section of the plane of the Milky Way with vertical relief showing the brightness of hydrogen. The local arm appears like a range of mountains on the right of the plot, with the Perseus arm curving out to its left. The more distant outer arm can just be seen appearing like 'foothills' even farther to the left.

The fact that our Galaxy has a spiral structure is somewhat of a puzzle (Figure 8.11). In its life, our Sun has circled the galactic centre about 20 times so why have the spiral arms not wound up? The solution is hinted at by a visual clue. Spiral arms seen in other galaxies stand out because they contain many bright blue stars – remember a single very hot star can outshine 50 000 suns like ours! However, very hot bright stars must be young as they have very short lives, so the spiral structure we see now is not that which would have been observed in the past. As Bertil Lindblad first suggested, it appears that the spi-ral arms are transitory and caused by a spiral density wave rotating round the galactic centre – a ripple that sweeps around the galaxy moving through the dust and gas. This compresses the gas as it passes and can trigger the collapse of gas clouds so forming the massive blue stars that delineate the spiral arms. The young blue stars show us where the density wave has just passed through but

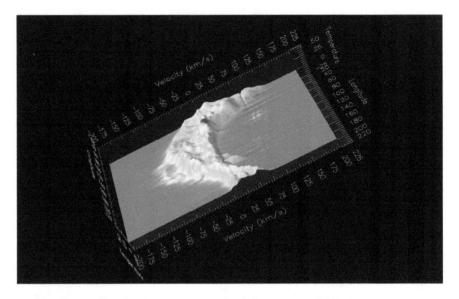

**Figure 8.10**  A relief map of a section of the Milky Way obtained with the 6.4-m radio telescope at Jodrell Bank Observatory. Image: Tim O'Brien, University of Manchester.

**Figure 8.11**  The spiral structure of our Galaxy as shown by observations of the hydrogen line.

in its wake will be left myriads of longer lived (and less bright) stars that form a more uniform disc.

### 8.1.6    *A super-massive black hole at the heart of our galaxy*

Whereas at optical wavelengths the centre of our Galaxy is obscured by dust, at radio wavelengths we are able to peer deep into its heart and astronomers have discovered a very compact radio source called Sgr A* in the constellation Sagittarius which we believe marks the position of a super-massive black hole at the centre of our Galaxy. How can we be sure that this exists? In the same way that we calculated the mass of the Sun by knowledge of the orbital velocity of the Earth and its distance from the Sun, we can estimate the mass of Sgr A* by measuring the speeds of stars in orbit around it at very close distances. For example, one of the 8-m VLT telescopes at Paranal Observatory in Chile has observed a star in the infrared as it passed just 17 light hours from the centre of the Milky Way (three times the distance of Pluto from the Sun). This convincingly showed that it was under the gravitational influence of an object that had an enormous gravitational field yet must be extremely compact – a super-massive black hole. Its mass is now thought to lie between 3.2 and 4 million solar masses confined within a volume one-tenth the size of the Earth's orbit.

## 8.2    Other galaxies

Galaxies, called originally 'white nebulae' have been observed for hundreds of years, but it was not until the early part of the last century that the debate as to whether they were within or beyond our Galaxy was settled – essentially when observations of Cepheid variables enabled their distances to be measured. They are, of course, objects outside our Galaxy and can now be observed throughout the universe. Galaxies can be divided into a number of types and then subdivided further to produce a classification scheme first devised by Edwin Hubble. As more and more galaxies were discovered, it became apparent that galaxies form groups (up to about ~100 galaxies) or clusters (containing hundreds to thousands of galaxies).

### 8.2.1    *Elliptical galaxies*

These, as their name implies, have an ellipsoidal form rather like a rugby ball (Figure 8.12). They range from those that are virtually circular in observed shape, called E0 by Hubble, to those, called E7, which are highly elongated (Figure 8.13).

**Figure 8.12** The giant elliptical galaxy, ESO 325-G004, in the Abell cluster S0740. Image: J. Blakeslee (Washington State University), NASA, ESA, and The Hubble Heritage Team (STScI/AURA).

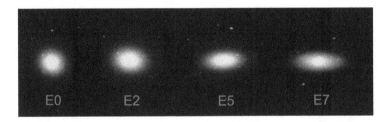

**Figure 8.13** Classification of elliptical galaxies.

This is a purely observational classification and does not necessarily tell us how elongated a galaxy really is – if seen end on even a highly elongated galaxy or rugby ball will look circular! At the heart of large galaxy clusters one or more giant elliptical galaxies are often observed. They may contain up to 10 million million solar masses in a volume some nine times that of our own Galaxy ~300 000 light years across. They are probably the result of the merger of many smaller galaxies. These are the most massive of all galaxies but are comparatively rare. Far more common are elliptical galaxies containing perhaps a few million solar masses within a volume a few thousand light years across. One interesting fact is that ellipsoidal galaxies do not appear to have any young stars within them so star formation appears to have ceased – all the gas having been used up to form stars in the past. Photographs of M31, the Andromeda Nebula, show its two companion elliptical galaxies. The nearer, more spherical in appearance, is of Type E2, whilst the more distant is of Type E5. Elliptical galaxies account for about one-third of all galaxies in the universe.

## 8.2.2    *Spiral galaxies*

Like our own Galaxy, these have a flattened spiral structure. The first observation of the spiral arms in a galaxy was made by the Third Earl of Rosse. During the 1840s he designed and had built the mirrors, tube and mountings for a 72 in. reflecting telescope which for three-quarters of a century was the largest optical telescope in the world. With this instrument, situated at Birr Castle in Ireland, Lord Rosse made some beautiful drawings of astronomical objects. Perhaps the most notable was that, shown in Figure 8.14, of an object which was the 51st to

**Figure 8.14** M51, the Whirlpool Galaxy, as drawn by the Third Earl of Rosse using the 72 in. telescope at Birr Castle in County Offaly, Ireland.

**Figure 8.15** Hubble Space Telescope image of M51. Image: NASA, ESA, S. Beckwith (STScI), and The Hubble Heritage Team (STScI/AURA).

be listed in Messier's catalogue and known as the 'Whirlpool Galaxy'. It was the first drawing to show the spiral arms of a galaxy and bears excellent comparison with modern day photographic images (Figure 8.15). (The galaxy is interacting with a second galaxy, NGC 5195; shown in the lower left of Figure 8.14.)

Spiral galaxies make up the majority of the brighter galaxies. Hubble classified them first into four types: S0, Sa, Sb and Sc. S0 galaxies, often called 'lenticular' galaxies, have a very large nucleus with hardly visible, very tightly wound, spiral arms. As one moves towards Type Sc, the nucleus becomes relatively smaller and the arms more open. In many galaxies the spiral arms appear to extend from either end of a central bar. These are called 'barred spirals' and are denoted SBa, SBb and SBc. Our own Milky Way Galaxy was thought to be a Sb or Sc galaxy but there is now some evidence that it has a bar making it an SBb or SBc galaxy. Type S0 galaxies, like elliptical galaxies, do not appear to have star forming regions, but as one moves towards the Type Sc or SBc, which contain more gas, star forming regions and the resulting young O and B type stars are seen in abundance.

Colour images show a significant colour contrast between the central core and the spiral arms. The core has a yellow-orange tint, whilst the spiral arms are bluish. The colour of the core indicates that the stars there are old, as yellow, orange and red stars live for long periods of time. In contrast, the spiral arms contain

**Figure 8.16** M81 in Ursa Major. Image: NASA, ESA, and The Hubble Heritage Team (STScI/AURA).

many young blue stars. Though in percentage terms they are relatively rare they outshine stars like our Sun by thousands of times so their light tends to dominate the colour of the spiral arms. Though the centre of the Galaxy M81, shown in Figure 8.16, is somewhat overexposed, the colour differences can be seen in the image as the colours have been slightly enhanced.

### 8.2.3     *Evidence for an unseen component in spiral galaxies – dark matter*

In the 1970s a problem related to the dynamics of galaxies came to light. Vera Rubin observed the light from HII regions (ionized clouds of hydrogen such as the Orion Nebula) in a number of spiral galaxies. These HII regions move with the stars and other visible matter in the galaxies but, as they are very bright, are easier to observe than other visible matter. HII regions emit the deep red hydrogen alpha (H-alpha) spectral line. By measuring the Doppler shift in this spectral line Rubin was able to plot their velocities around the galactic centre as a function of their distance from it. She had expected that clouds that were more distant from the centre of the galaxy (where much of its mass was expected to be concentrated) would rotate at lower speeds – just as the outer planets travel more slowly around the Sun. This is known as Keplerian motion, with the rotational speed decreasing inversely as the square root of the distance from the centre. (This is enshrined in Kepler's third law of planetary motion and can be derived from Newton's law of gravity.)

To her great surprise, Rubin found that the rotational speeds of the clouds did not decrease with increasing distance from the galactic centre and, in some cases, even increased somewhat. Not all the mass of the galaxy is located in the centre but the rotational speed would still be expected to decrease with increasing radius beyond the inner regions of the galaxy although the decrease would not be as rapid as if all the mass were located in the centre. To give an example; the rotation speed of our own Sun around the centre of the Milky Way Galaxy would be expected to be ~160 km s$^{-1}$. It is, in fact, ~220 km s$^{-1}$. The only way these results can be explained is that either the stars in the galaxy are embedded in a large halo of unseen matter – extending well beyond the visible galaxy – or that Newton's law of gravity does not hold true for large distances. The unseen matter, whose gravitational effects Rubin's observations had discovered, is called 'dark matter'. Further evidence, some dating from the 1930s and some very recent, for the existence of dark matter will be given below. A more detailed discussion will be given in Chapter 9.

A modified form of Newton's law called MOND (MOdified Newtonian Dynamics) was proposed by Mordechai Milgrom in 1981, who pointed out that Newton's Second Law ($F = ma$) when applied to gravitational forces has only been verified when the gravitational acceleration is large and has never been verified where the acceleration, $a$, is extremely small – as would be the case for stars towards the edge of a galaxy where the gravitational forces are very weak. With a suitable choice of parameters the observed rotation curves of galaxies can be accurately modelled by the MOND theory (Figure 8.17); however, it has a much harder task explaining other observations that support the existence of dark matter, such as the dynamics of galaxy clusters and gravitational lensing, so MOND will not be considered further.

### 8.2.4    *Weighing a galaxy*

The observations of the hydrogen line described above can be used to calculate the mass of a galaxy. Figure 8.18 shows the hydrogen line spectrum of the nearby galaxy M33 which lies at a distance of $2.36 \times 10^{22}$ m (~2.9 million light years).

The horizontal axis of Figure 8.18 has been converted from frequency to velocity using the Doppler formula: $\Delta f / f = v/c$. The hydrogen line spectrum of M33 has a width in frequency due to the fact that it is rotating – one side of the galaxy is coming towards us whilst the other is moving away. The centre of the M33 hydrogen emission corresponds to a velocity of $-180$ km s$^{-1}$. You might well deduce that the galaxy as a whole is moving *away* from us at this speed but, as all but a few galaxies are moving away from us, the sign convention that is used is that galaxies moving away from us are given positive velocities and those moving

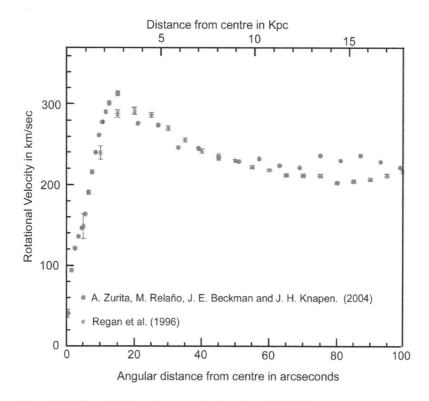

**Figure 8.17** The galactic rotation curve for the galaxy NGC 1530.

towards us are given negative velocities. So this indicates that M33 is moving *towards* us at a speed of ~180 km s$^{-1}$. However, our Solar System is moving around the centre of the galaxy at a speed of ~220 km s$^{-1}$ and, having corrected for this, M33 is actually moving towards the Milky Way Galaxy at a speed of ~24 km s$^{-1}$.

The width of the spectral line is ~200 km s$^{-1}$ so that the hydrogen at the edge of the galaxy is apparently moving around the centre at a speed of 100 km s$^{-1}$. However, though the galaxy is presumably circular, its dimensions on a photographic plate are ~71 × 45 arcmin. This implies that it is inclined to our line of sight at an angle of arcsin (45/71) = ~39°. As a result, the value we measure will be less than the true value due to the projection effect. (If the galaxy was perpendicular to us, we would not observe any rotational width in the hydrogen line spectrum.) The true rotational velocity of the outer parts of the galaxy about its centre should thus be close to 100/sin(39) = 158 km s$^{-1}$.

Knowing the distance of the galaxy and its angular size we can calculate its radius.

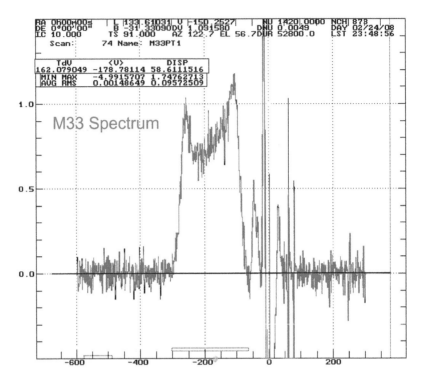

**Figure 8.18** The hydrogen line spectrum of M33 in Triangulum. Image: Christine Jordan, University of Manchester.

M33 is ~71 arcmin across which is $71/(60 \times 57.3) = 0.020$ rad. It lies at a distance of $2.36 \times 10^{22}$ m.

The radius of M33 is thus ~$0.5 \times 0.020 \times 2.36 \times 10^{22}$ m
$$= \sim 2.4 \times 10^{20} \text{ m}.$$

If the mass distribution of the galaxy is symmetrical then the gravitational effect on the hydrogen gas at the edge of the galaxy is the same as if all of the galaxy's mass was concentrated at its centre. One can thus use an identical method to that used to calculate the mass of the Sun.

The gravitational force on a small mass at this distance must equal the centripetal acceleration:

$$G M m/r^2 = m\, v^2/r$$

where $M$ is the mass of the galaxy, $m$ is the mass of a small volume of hydrogen, $r$ is the distance of the hydrogen from the centre of the galaxy and $v$ is the velocity of hydrogen around the centre.

This gives:

$$M = r v^2 / G$$
$$= 2.4 \times 10^{20} \times (1.58 \times 10^5)^2 / 6.67 \times 10^{-11} \, \text{kg}$$
$$= 9 \times 10^{40} \, \text{kg}$$
$$= 9 \times 10^{40} / 2 \times 10^{30} \, \text{solar masses}$$
$$= \sim 45 \, \text{thousand million solar masses.}$$

We have another method of estimating the mass using what is termed the 'mass to light ratio' of stars. This is simply the ratio between the mass of a star or star cluster divided by its luminosity – our Sun has, by definition, a mass of 1 and a luminosity of 1 so its mass to light ratio is 1. One could assume that all the stars in M33 were similar to our Sun in terms of their mass to light ratio. If we then calculate the luminosity of M33 compared with that of our Sun we will directly get an estimate of the mass of M33 in solar masses.

The absolute magnitude of the Sun is 4.8 and that of M33 is $-19.5$, a difference of 24.3 magnitudes. This corresponds to a difference in luminosity of $2.512^{24.3} = \sim 5.2 \times 10^9$, which gives a mass estimate of $\sim 5$ thousand million solar masses. This is a factor of 10 less than the value derived above. M33 obviously has mass which does not emit light, such as dust and gas, and not all stars will have the same mass to light ratio as our Sun—hot stars are very luminous for their mass compared with our Sun and cooler stars (of which there are many more) less luminous. The average mass to light ratio for stars, gas and dust in our own Galaxy is $\sim 1.5$ so, assuming a similar mix, this would give M33 a mass of $\sim 8$ thousand million solar masses.

The fact that this is still a factor of $\sim 6$ less than that derived dynamically is further evidence of the presence of dark matter in the galaxy – there appears to be $\sim 5$ times as much dark matter than normal matter in the makeup of the galaxy!

## 8.2.5    *Irregular galaxies*

A small percentage of galaxies show no obvious form and are classified as Irregulars. One nearby example is the Small Magellanic Cloud (SMC). Its companion, the Large Magellanic Cloud (LMC), is usually classified as one too, though it shows some features of a small barred spiral. Such small galaxies are not very bright so we cannot see too many but they may, in fact, be the most common type. They contain enough gas to allow star formation, but relatively less dust than found in our Galaxy. In 30 Doradus, often called the Tarantula Nebula because of its spidery appearance, the LMC has one of the largest star formation regions known. It contains many young stars; one of these gave rise to a supernova, called 1987A, observed in February 1987 – the nearest to us for several hundred years.

### 8.2.6    *The Hubble classification of galaxies*

A schematic, called 'Hubble's Tuning Fork' is used to illustrate Edwin Hubble's classification scheme (Figure 8.19). Hubble believed that, as they aged, galaxies evolved from the left end of the tuning fork toward the right. He therefore called elliptical galaxies 'early galaxies' and spiral galaxies 'late galaxies'. This idea was not correct. Spiral galaxies rotate, and so have angular momentum, whilst elliptical galaxies do not. There is no way that elliptical galaxies could suddenly gain angular momentum so they could not turn into spiral galaxies. It is, however, the case that the merging of spiral galaxies probably results in the formation of giant elliptical galaxies.

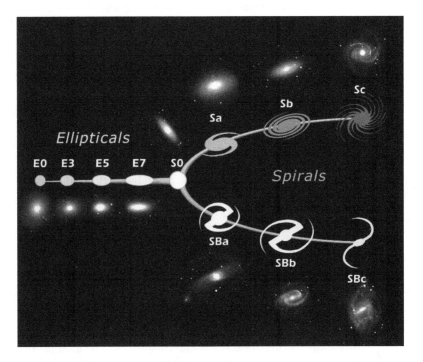

**Figure 8.19** Diagram summarizing the Hubble galaxy classification scheme. Image: NASA, Wikipedia Commons.

## 8.3    The universe

As our telescopes look ever further out into space, we say that we are observing our universe. Assuming that the universe has an origin some time in the past, there is a limit to how far we can see due to the finite speed of light – we cannot see further than the distance light can travel since its origin. As the universe ages, this limit increases. The part of the universe that we *can* see is properly called the 'visible universe', but this is normally shortened to simply the 'universe'. The whole of space, of which our universe is a part, may well extend far beyond – possibly to infinity – and the totality of space is often termed the 'cosmos', hence the term 'cosmology' used to define the study of its origin and evolution.

### 8.3.1    *The cosmic distance scale*

To understand the scale size of the universe we need to be able to measure the distances of the galaxies. Key reference points in the 'distance ladder' are the distances to the Large and Small Magellanic Clouds, two nearby irregular galaxies (Figure 8.1). As we will see, a type of very bright variable star observed within them, the Cepheid variables, can be used as a 'standard candle' to measure the distances of galaxies relative to the distances of the Magellanic Clouds. For this to be possible, we must first have an accurate measure of the distances of one or other of the Magellanic Clouds.

Estimating the precise distances to the LMC and SMC has not been easy, with values varying substantially over the years. With the occurrence of Supernova 1987A in the LMC, astronomers were given a direct method to measure the distance to the LMC. This was based on the time taken for the light emitted by the supernova to illuminate a distant ring of gas surrounding it. These important observations and the calculations based on them are described in slightly simplified form below and have resulted in an estimate of the distance from Earth to the centre of the LMC of $52.0 \pm 1.3$ kpc. This, together with other observations that are in agreement, has helped define a new value for the zero point of the Cepheid distance scale, and has greatly improved our knowledge of galactic distances.

### 8.3.2    *Using Supernova 1987A to measure the distance of the Large Magellanic Cloud*

Some time after the supernova was first observed, a ring of material that had previously been ejected by the progenitor star lit up as it was irradiated by ultraviolet light from the explosion. Had the ring been at right angles to our line of sight (Figure 8.20a) we would have seen all parts light up simultaneously and, as it

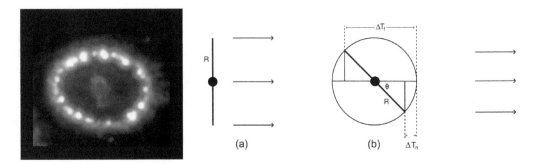

**Figure 8.20** The geometry of the ring surrounding Supernova 1987A. Image: NASA, P. Challis, R. Kirshner (Harvard-Smithsonian Center for Astrophysics) and B. Sugerman (STScI).

would then have been at the same distance as the supernova, the number of days this was after we first saw the supernova would simply be the radius of the ring in light days, $R$. However, as the ring is at an angle $\theta$ to our line of sight, we saw the nearest part of the ring brighten after 75 days and the farthest part of the ring brighten after 390 days. From Figure 8.20b it can be seen that we would have observed the nearest part of the ring brighten after a time $\Delta T_n$ after we first saw the supernova and the most distant part brighten after a time $\Delta T_f$.

From the geometry and the observations one can show that it took 232.5 days for light to reach the ring, a distance $R$ from the supernova, after the explosion.

Assume that the distance $R$ is measured in light days, then:

The light from the near side of the ring will be seen to light up a time:

$$\Delta T_n = (R - R\cos\theta) \text{ days after we first see the supernova.}$$

The light from the far side of the ring will be seen to light up a time:

$$\Delta T_f = (R + R\cos\theta) \text{ days after we first see the supernova.}$$

Adding these two equations gives:

$$\Delta T_n + \Delta T_f = [(R - R\cos\theta) + (R + R\cos\theta)]$$
$$= 2R$$

So
$$R = (\Delta T_n + \Delta T_f)/2 = (75 + 390)/2$$
$$= 232.5 \text{ light days.}$$

The Hubble Space Telescope (HST) image gave an angular size for the radius of the ring of 0.86 arcsec. This angular size and the deduced ring radius of 232.5 light days can be used to estimate the distance of the LMC in parsecs.

$$\text{Radius of the ring} = 232.5 \times 3 \times 10^5 \times 3600 \times 24\,\text{km}$$
$$= 6 \times 10^{12}\,\text{km}$$
$$\text{(The velocity of light is } 3 \times 10^5\,\text{km s}^{-1} \text{ and } 1\,\text{pc} = 3.1 \times 10^{13}\,\text{km.)}$$

The semi-major axis of the ellipse observed by the HST will be the radius of the ring. (It is an ellipse as the ring is seen at an angle to the line of sight.)

The ring radius thus subtends an angle of 0.86 arcsec, so the distance, $D$, will be given by:

$$D = d/\theta \text{ (where } \theta \text{ is in radians)}$$
$$= 6 \times 10^{12}/[0.86/(3600 \times 57.3)]\,\text{km}$$
$$= 6 \times 10^{12}/4.2 \times 10^{-6}\,\text{km}$$
$$= 1.4 \times 10^{18}\,\text{km}$$
$$= 1.4 \times 10^{18}/3.1 \times 10^{13}\,\text{pc}$$
$$= 46\,100\,\text{pc}$$

This is a somewhat lower than the 'accepted' value of ~50 000 pc. The discrepancy is due to the fact that the ring has a thickness and its actual size is perhaps 7% greater than that calculated by the times that the near and far sides of the ring were seen to brighten (which would correspond to when the *inner diameter* of the ring would be illuminated by ultraviolet light rather than the *outer diameter* of the ring as measure by the HST). A recent very detailed analysis of the light curves which allow the ring radius to be measured more accurately give a value of the distance of Supernova 1987A of 51.5 +/− 1.2 kpc.

## 8.3.3     The Cepheid variable distance scale

The observational basis of this distance scale was provided by Henrietta Leavitt whilst working at the Harvard College Observatory where she became head of the photographic photometry department. In the early years of the twentieth century, her group studied images of stars to determine their magnitude using a photographic measurement system developed by Leavitt that covered a 17 magnitude brightness range. Many of the plates measured by Leavitt were taken at Harvard Observatory's southern station in Arequipa, Peru from which the Magellanic Clouds could be observed and she spent much time searching the plates taken there for variable stars within them. She discovered many variable stars within them including 25 Cepheid variable stars. These stars are amongst some of the brightest; between 1000 and 10 000 times that of our Sun and are named after the star Delta Cephei which was discovered to be variable by the British astronomer John Goodricke in 1784. These stars pulsate regularly rising rapidly to peak brightness and falling more slowly, as shown in Figure 8.21a. Leavitt determined the periods of 25 Cepheid variables in the SMC and in 1912 announced what has since become known as the famous period–luminosity relationship (Figure 8.21b). She stated: 'A straight line can be readily drawn among each of the two series of points corresponding to maxima and minima (of the brightness of Cepheid variables), thus showing that there is a simple relation between the brightness of the variable and their periods.' As the SMC was at some considerable distance from Earth and was relatively small, Leavitt also realized that: 'as the variables are probably nearly the same distance from the Earth, their periods are apparently associated with their actual emission of light, as determined by their mass, density, and surface brightness.'

The relationship between a Cepheid variable's luminosity and period is quite precise, and a 3-day period Cepheid corresponds to a luminosity of about 800 times the Sun whilst a 30-day period Cepheid is 10 000 times as bright as the Sun. To give an example; we might measure the period of a Cepheid variable in a distant galaxy and observe that it is 1000 times fainter than a Cepheid variable in the LMC. We can thus deduce that, from the inverse square law, it would be 100 times further away than the LMC, that is $100 \times 51.2$ kpc giving a distance of 5100 kpc (16 600 Ly). Cepheid stars are thus the ideal standard candle to measure the distance of clusters and external galaxies. (As we do not know the precise location of the Cepheid variable within the cluster or galaxy there will be a small uncertainty but this error is typically small enough to be irrelevant.)

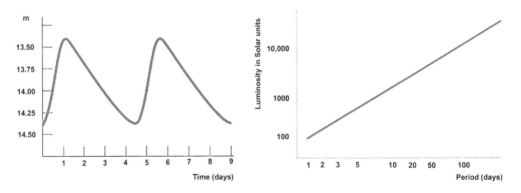

**Figure 8.21** (a) A Cepheid variable light curve and (b) the period–luminosity relationship.

Edwin Hubble, whose famous work based on his measurements of the distances of galaxies is discussed in Chapter 9, identified some Cepheid variables in the Andromeda Galaxy, proving for the first time its extragalactic nature. Until this time some astronomers had argued that the so-called 'white nebulae' were within our own Milky Way Galaxy. Recently, the HST succeeded in identifying some Cepheid variables in the galaxies of the Virgo cluster, so enabling a determination of its distance of 18 Mpc (60 million Ly).

### Example: to measure the distance of M81 using Cepheid variable data

M81 is shown in Figure 8.16.

From the plots in Figure 8.22, a Cepheid variable, the log of whose period is 1.4 (~25 days), will have a visual magnitude of 13.9 in the LMC and 22.8 in M81 – a difference of 8.9 magnitudes.

This corresponds to a brightness ratio of $2.512^{8.9} = 3632$. Using the inverse square law this means that M81 must be $(3632)^{1/2} = 60.2$ times further away than the LMC.

Given the distance of the LMC deduced from observations of Supernova 1987A this gives a distance of M81 of $51.5 \times 60.2\,\mathrm{kpc} = 3104\,\mathrm{kpc} = 10.1$ million Ly.

### 8.3.4     *Starburst galaxies*

Starburst galaxies emit more than usual amounts of infrared light and radio waves and came to prominence when surveys of the sky were made with infrared

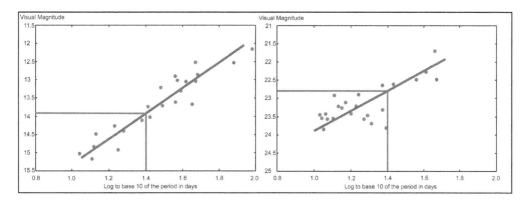

**Figure 8.22** Plots of visual magnitude against $\log_{10}$ of the period for Cepheid variables in the Large Magellanic Cloud (a) and M81 (b).

**Figure 8.23** M82 in Ursa Major – a starburst galaxy. Image: J. Gallagher (University of Wisconsin), M. Mountain (STScI), and P. Puxley (National Science Foundation), NASA, ESA, and The Hubble Heritage Team (STScI/AURA).

telescopes. A nearby example is M82, 12 million light years away in Ursa Major (Figure 8.23). It appears that a close passage with M81, its neighbour in space, has triggered a rapid burst of star formation. The radiation from young stars heats up the dust in the galaxy so producing the infrared emission. Sometimes, where there is less dust, ultraviolet light from the very hot stars is seen too. The HII regions where the stars are forming emit strongly in the radio part of the spectrum so helping to produce the enhanced radio emission. Amongst the stars that are born will be a small number of very massive stars that evolve quickly and then die in spectacular supernova explosions. Further radio emission then comes from the electrons accelerated to speeds close to that of light in the explosions that mark their death. In the heart of M82, hidden from optical telescopes by thick dust lanes, radio telescopes have located over 50 supernova remnants. Using an array of telescopes stretching across Europe some of these have been imaged. Comparing images made in 1986 and 1997 astronomers found that their shells of gas were expanding at speeds of up to $20\,000\,\mathrm{km\,s^{-1}}$. The youngest supernova exploded about 35 years ago but, hidden behind a curtain of dust, this was never seen by optical telescopes.

### 8.3.5    *Active galaxies*

These are galaxies where some processes occurring within them make them stand out from the normal run of galaxies, often in the amount of radio emission that they produce. It was mentioned earlier that at the heart of our Galaxy lies a radio source called Sgr A*, one of the strongest radio sources in our Galaxy. However, this would be too weak to be seen if our Milky Way Galaxy were at a great distance and our Galaxy would therefore be termed a 'normal' galaxy. However, there are some galaxies that emit vastly more radio emission and shine like beacons across the universe. As most of the excess emission lies in the radio part of the spectrum, these are called radio galaxies. Other galaxies produce an excess of X-ray emission and, collectively, all are called active galaxies. Though relatively rare, there are obviously energetic processes going on within them that make them interesting objects for astronomers to study.

We believe that the cause of their bright emissions lies right at their heart in what is called an active galactic nucleus (AGN). It was mentioned earlier that we believe that there is a super-massive black hole at the centre of our Galaxy. We now believe that black holes, containing up to several billion solar masses, exist at the centre of all large elliptical and spiral galaxies. In the great majority of galaxies these are quiescent but in some, matter is currently falling into the black hole fuelling the processes that give rise to the X-ray and radio emission.

As an example, the Faint Object Spectrograph aboard the HST has observed gas in a rotating disc at a distance of ~26 Ly from the galactic centre of M84. At this distance on one side of the centre, the gas is moving towards us at speed of about $400\,\mathrm{km\,s^{-1}}$ (relative to the centre of the galaxy) whilst, at the same distance on other side, it is moving away from us at the same speed.

From these data we can compute the mass of the central region using exactly the same calculation as was used to compute the mass of M33:

$$M = r\,v^2/G$$

where $M$ is the mass of the central region of the galaxy, $r$ is the distance of the gas from the centre of the galaxy and $v$ is the velocity of the gas around the centre.

$$26\,\mathrm{Ly} = 26 \times 9.46 \times 10^{15}\,\mathrm{m} = 2.4 \times 10^{17}\,\mathrm{m},\; v = 4.0 \times 10^5\,\mathrm{m\,s^{-1}}$$

This gives:
$$M = 2.4 \times 10^{17} \times (4 \times 10^5)^2/6.67 \times 10^{-11}\,\mathrm{kg}$$
$$= 5.9 \times 10^{38}\,\mathrm{kg}$$
$$= 5.9 \times 10^{38}/2 \times 10^{30}\,M_{sun}$$
$$= 2.93 \times 10^8\,M_{sun}$$
$$= {\sim}300 \text{ million } M_{sun}$$

It is expected that most of this mass will be in a black hole near the centre of the galaxy. The same calculation as was used to find the Schwarzschild radius ($R_S$) for a stellar mass black hole in Chapter 7 can be used to find the approximate size of this super-massive black hole:

$$R_S = 3.0 \times M/M_{sun} \text{ km}$$
$$= 3.0 \times 5.9 \times 10^{38}/2 \times 10^{30} \text{ km}$$
$$= 8.8 \times 10^8 \text{ km}$$

This is somewhat less than the size of the orbit of the planet Venus.

Let us consider what happens as a star begins to fall in towards the black hole. As one side will be closer to the black hole than the other, the gravitational pull on that side will be greater than on the further side. This exerts a differential gravitational force on the star, called a tidal force, which increases as the star gets closer to the black hole. The final effect of this tidal force will be to break the star up into its constituent gas and dust. It is unlikely that a star would be falling in directly towards the black hole and would thus have some rotational motion – that is, it would be circling around the black hole as well as gradually falling in towards it. As the material gets closer it has to conserve angular momentum and so speeds up – just like ice skaters bringing their arms in towards themselves. The result of the material rotating round in close proximity at differing speeds is to produce friction so generating heat that causes the gas and dust to reach very high temperatures. The black hole region is surrounded by a torus (or doughnut) of material called an accretion disc that contains so much dust that it is opaque. However, if, by chance, this torus lies roughly at right angles to our line of sight then we can see in towards the black hole region and will observe the emissions coming from the material.

Nuclear fusion of hydrogen can convert just under 1% of its rest mass into energy. What is less obvious is that the act of falling into a gravitational potential well can also convert mass into energy. In the case of a super-massive black hole energy equivalent to at least 10% of the mass can be released before it falls within the event horizon – giving the most efficient source of energy that we know of! If the black hole is rotating, the energy release can be even higher. This often results in the formation of two opposing jets of particles moving away from the black hole along its rotation axis. Moving at speeds close to that of light, these 'bore' a hole through the gas surrounding the galaxy and in doing so the particles will be slowed down or decelerated. They then produce radiation across the whole electromagnetic spectrum and thus allow the jets to be observed. If one of the jets happens to be pointing towards us, the observed emission can be very great and so these objects can be seen right across the universe.

These highly luminous objects were first discovered by radio astronomers in a series of experiments designed to measure the angular sizes of radio sources. In the early 1960s, the signals received with the 75-m Mark I radio telescope at Jodrell Bank were combined with those from smaller telescopes located at increasingly greater distances across the north of England. The observations showed that a number of the most powerful radio sources had angular sizes of less than 1 arcsec and would appear as 'stars' on a photographic plate. They were thus given the name 'quasi-stellar-object' (looking like a star) or 'quasar' for short. This meant that they were very hard to identify until their precise positions were known. The first quasar to be identified was the 273rd object in the Third Cambridge catalogue of radio sources so it was given the name 3C273.

Though its image, shown in Figure 8.24, taken by the 5-m Hale telescope, looked very like a star, a jet was seen extending ~6 arcsec to one side. It was discovered that its distance was about 611 Mpc or 2000 million light years – then the most distant object known in the Universe. However, 3C273 is one of the closer quasars to us and the most distant currently known lies at a distance of ~4000 Mpc, or 13 billion light years! Hence, quasars are some of the most distant and most luminous objects that can be observed in the Universe.

An interesting exercise is to calculate how much mass a quasar must 'consume' in order to give its observed brightness. If we assume that 10% of the mass is

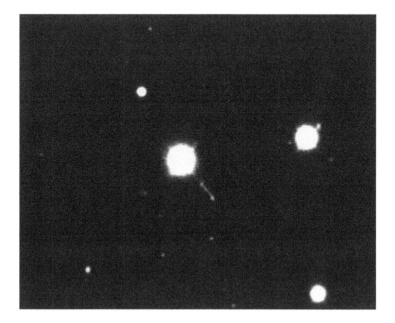

**Figure 8.24** 3C273 as imaged by the 5-m Hale Telescope on Mount Palomar.

converted into energy, then $E = (1/10)mc^2$ giving $m = 10E/c^2$. The brightest quasars have luminosities of order $10^{41}$ W. (This is $10^{41}$ J s$^{-1}$, so we must use $3 \times 10^8$ m s$^{-1}$ for our value of $c$.) This equation will then give the mass required per second:

$$m_s = 10 \times 10^{41}/(3 \times 10^8)^2 \, \text{kg}$$
$$= 1.1 \times 10^{25} \, \text{kg}$$

So, the mass per year:    $m_{year} = 86400 \times 365 \times 1.1 \times 10^{25} \, \text{kg}$
$$= 3.5 \times 10^{32} \, \text{kg}$$

As usual, this can be converted into solar masses:

$$= 3.5 \times 10^{32}/2 \times 10^{30}$$
$$= \sim\!175 \text{ solar masses per year}$$

The Sloan Digital Sky Survey (SDSS) has observed more than 120 000 galaxies. Of these, 20 000 contain massive black holes in their centres that are currently active and growing in mass – showing that many galaxies have super-massive black holes at their centre. Whether nuclei of galaxies are 'active' is thus a question of whether mass is being fed into these black holes.

### The size of the active galactic nucleus

There is a simple observation of quasars that can give us an indication of the size of the emitting region around the black hole. It has been observed that the light and radio output of a quasar can change significantly over periods of just a few hours. Perhaps surprisingly, this can provide a reasonable estimate of its size as the following 'thought experiment' will show.

Suppose the Sun's surface instantly became dark. We would see no change for 8.32 min due to the time the light takes to travel from the Sun to the Earth. Then we would first see the central region of the disc go dark as this is nearest to us, and the light travel time from it is least. This dark region would then be seen to expand to cover the whole of the Sun's visible surface. This is because the light from regions of the Sun further from us would still be arriving after the light from the central region was extinguished. The time for the change to occur would be given by: $t = r_{sun}/c$.

The radius of the Sun is 695 000 km so the time for the whole of the Sun to darken would be given by:

$$695\,000/3 \times 10^5 \, \text{s} = 2.31 \, \text{s}.$$

It is thus apparent that a body cannot appear to instantaneously change its brightness and can only do so on timescales of order of the light travel time across the radiating body.

Suppose that an AGN is observed to significantly change its brightness over a period of 12 h:

$$12\,h \text{ is } 12 \times 60\,min = 720\,min.$$

As light can travel 1 AU in 8.32 min the scale size of the object must be of order $720/8.32\,AU = {\sim}86\,AU$.

### 8.3.6    *Groups and clusters of galaxies*

Most galaxies are found in groups typically containing a few tens of galaxies or clusters that may contain up to several thousand. Our Milky Way Galaxy forms part of the Local Group that contains around 40 galaxies within a volume of space 3 million light years across. Our Galaxy is one of the three spiral galaxies (along with M31 and M33) that dominate the group and contain the majority of its mass. M31, the Andromeda Galaxy, shown in Figure 8.25, and our own are comparable in size and mass and their mutual gravitational attraction is bringing them towards each other so that, in a few billion years, they will merge to form an elliptical galaxy. M33, whose mass we calculated earlier, is the third largest galaxy in the group. The group also contains many dwarf elliptical galaxies, such as the two orbiting M31 and visible in Figure 8.24. M32 is a small, Type E2, elliptical galaxy seen to the left of the nucleus of M31, just outside its spiral arms, whilst NGC 205 (M110) is a more elongated, Type E5 or E6, elliptical that is seen to the lower left of the nucleus of M31. The Local Group contains several large irregular galaxies, such as the Magellanic Clouds, and at least 10 dwarf irregulars to add to the total. There may well be more galaxies within the group, hidden beyond the Milky Way that obscures over 20% of the heavens.

The region lying in Virgo just to the west of Leo shows many hundreds of galaxies and is thus called 'the realm of the Galaxies'. Sixteen of these are bright enough to have been catalogued by Charles Messier using a telescope with an aperture of just a few inches. In this direction we are looking towards the heart of a galaxy cluster containing some 2000 members called, due the constellation in which we observe it, the Virgo cluster (Figure 8.26). Two other nearby clusters are the Coma cluster and the Hercules cluster. Galaxy clusters typically contain 50–1000 galaxies within diameters of 6–35 million light years and having total masses of $10^{14}$–$10^{15}$ solar masses.

**Figure 8.25** M31, the Andromeda Galaxy, along with two dwarf elliptical galaxies, M32 and NGC 205. Image: Adam Block, NOAO/AURA/NSF.

**Figure 8.26** The Virgo cluster of galaxies. Image: Space Telescope Science Institute, NASA.

### 8.3.7        *Superclusters*

Small clusters and groups of galaxies appear to make up structures on an even larger scale. They are known as superclusters and have overall sizes of order 300 million light years (100 times the scale size of our Local Group). Usually a supercluster is dominated by one very rich cluster surrounded by a number of smaller groups. The local supercluster is dominated by the Virgo cluster. The Virgo supercluster, as it is often called, is in the form of a flattened ellipse about 150 million light years in extent with the Virgo cluster at its centre and our Local Group near one end. In the same general direction but further away lies the Coma cluster containing over 1000 galaxies. It is the dominant cluster in the Coma supercluster at a distance of 330 million light years. Two other nearby superclusters lie in the directions of the constellations Perseus/Pisces and Hydra/Centaurus at distances of 150 and 230 million light years, respectively.

These very great distances are beyond that where the Cepheid variable distance scale can be used. In Chapter 9 we will see how relatively recent observations of Type Ia supernovae are being used to measure distances of this magnitude. However, it is worth mentioning a very simple method that had been used prior to this new technique. One could reasonably assume that the largest galaxies within a cluster are probably of the same size. One might take the third largest to avoid the possibility that the largest might be more extreme than in most clusters. If their sizes are measured on a photographic plate then one can compare them to a galaxy, hoped to be similar is size, whose distance we do know. In a similar vein, one could take the magnitude of the galaxy and use that to estimate its distance.

The most distant galaxies ever observed are in the Hubble Ultra Deep Field shown in Figure 8.27. Some are observed at a time just 0.7 billion years after the origin of the universe, just 5% of its present age.

### 8.3.8        *The structure of the universe*

Galaxies, in their groups, clusters and superclusters, extend across the visible universe, and recently it has been possible to map their distribution through space. Figure 8.28 shows a panoramic view of the entire sky images in the near infrared and reveals the distribution of galaxies beyond the Milky Way, shown in the centre of the image. The galaxies are colour coded by 'redshift'; in blue are the nearest galaxies, green are at moderate distances and red are the most distant.

The distribution is not uniform, but appears rather like the structure of a sponge. In analogy, the galaxies are located on the walls of the sponge and few are found within the empty spaces within the sponge. These nearly empty regions

**Figure 8.27** The Hubble Ultra Deep Field. Image: Beckwith and the HUDF Working Group (STScI), HST, ESA, NASA.

are called voids. How and why this structure has evolved over the life of the universe will be discussed in Chapter 9.

## 8.4    Questions

1.  Use 'Google' to find the website for the Digitized Sky Survey. It is possible to ask to observe a specific object and produce an image in Gif format. You might like to try this to produce an image of M81. First enter M81 into the 'Object name' box and click on 'Get coordinates'. Then enter 30 and 30 as the width and height – the units are in arcminutes. Select 'GIF' as the file format and finally click on 'Retrieve image'. An image of the region around M81 should then appear on your screen.

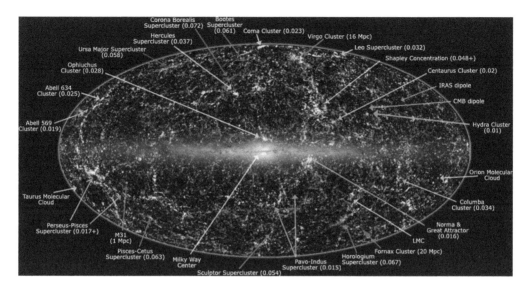

**Figure 8.28** The structure of the local universe. Image: T. Jarett (IPAC/Caltech), Wikipedia Commons.

The spiral galaxy M81 lies at a distance of 12 million light years. The width of its hydrogen line spectrum is $300\,\mathrm{km\,s^{-1}}$. From the image you have obtained, estimate the angular diameter of M81 and hence find its approximate radius in kilometres. Estimate its mass in solar masses following the same reasoning as given with regards to M33. (1 solar mass $= 2 \times 10^{30}\,\mathrm{kg}$, 1 light year $= 9.46 \times 10^{12}\,\mathrm{km}$, $G = 6.67 \times 10^{-11}\,\mathrm{m^3\,kg^{-1}\,s^{-2}}$.)

2.   Observations of the central region of the Galaxy M87 indicate that stars which are 60 light years from the centre are orbiting the central super-massive black hole at a speed of $550\,\mathrm{km\,s^{-1}}$. Estimate the mass of the black hole in solar masses. (The Earth orbits the Sun with an orbital speed of $30\,\mathrm{km\,s^{-1}}$. 1 light year is 63240 AU.)

3.   The emission from the region surrounding an accreting super-massive black hole is seen to vary significantly on the time scale of 12 h. Knowing that light takes 8.32 light min to travel from the Sun to the Earth, estimate the size of the region in astronomical units (AU).

4.   A Cepheid variable star is observed in a distant galaxy. It has a period of 100 days and a peak brightness 12 magnitudes less than the brightness of a 100-day period, Type II Cepheid variable observed in the Large Magellanic Cloud (LMC). Given that the LMC lies at a distance of 50 000 pc from our Sun, calculate the distance from our Sun to the distant galaxy.

## Chapter 9

# Cosmology – the Origin and Evolution of the Universe

## 9.1 Einstein's blunder?

Modern cosmology arose from Einstein's General Theory of Relativity which, as we have seen, is essentially a theory of gravity (Figure 9.1). As gravity was the only force of infinite range that could act on neutral matter, Einstein realised that the universe as a whole must obey its laws. He was led to believe that the universe was 'static', or unchanging with time, and this caused him a real problem as gravity, being an attractive force, would naturally cause stationary objects in space to collapse down to one point. To overcome this he had to introduce a term into his equations that he called the cosmological constant, lambda or $\Lambda$. This represents a form of antigravity that has the interesting property that its effects become greater with distance. So, with one force decreasing and the second increasing with distance it was possible to produce a static solution. He later realised that this was an unstable situation, and that a static universe was not possible, calling this 'the greatest blunder' of his life. He could have predicted that the universe must be either expanding or contracting. However, as we shall see, perhaps he was not as wrong as he thought.

## 9.2 Big Bang models of the universe

A Russian meteorologist, A.A. Friedmann, solved Einstein's equations to produce a set of models in which the universe expanded from a point, or singularity. These were given the name Big Bang models by Fred Hoyle – this was meant to be a disparaging term as Hoyle was an advocate of another theory, the Steady State theory, to be discussed below, and did not like them! In all of these models, the initially fast rate of expansion is slowed by the attractive gravitational force between the matter of the universe. If the density of matter within the universe exceeded a critical amount, it would be sufficient to cause the expansion to cease and then

*Introduction to Astronomy and Cosmology*   Ian Morison
© 2008 John Wiley & Sons, Ltd

**Figure 9.1** Albert Einstein in 1947. Image: Wikipedia Commons.

the universe would collapse down to a 'Big Crunch' (these are called 'closed' universes). If the actual density was less than the critical density, the universe would expand for ever (called 'open' universes). In the critical case that is the boundary between the open and closed universes, the rate of expansion would fall to zero after infinite time (called the 'flat' or 'critical' universe) (Figure 9.2).

A useful analogy is that of firing a rocket from the Earth; if the speed of the rocket is less than $11.186\,\mathrm{km\,s^{-1}}$, the Earth's escape velocity, the rocket will eventually come to a halt and fall back to Earth (equivalent to closed universes), if it equals $11.186\,\mathrm{km\,s^{-1}}$ it will escape the Earth with its speed reducing with time

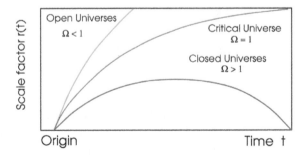

**Figure 9.2** The Friedmann models of an expanding universe.

(equivalent to the critical universe), whilst if it exceeds $11.186\,\mathrm{km\,s^{-1}}$ it will leave the Earth more quickly (equivalent to open universes).

The models are distinguished by a constant, $\Omega$ (omega) that is defined as the ratio of the actual density to the critical density. In closed universes, $\Omega$ is greater than 1, space has positive curvature, the angles within a triangle add up to more than 180° and two initially parallel light rays would converge. In open universes, $\Omega$ is less than 1, space has negative curvature, the angles within a triangle add up to less than 180° and two initially parallel light rays would diverge. In the critical case, $\Omega$ is equal to 1, space is said to be 'flat', the angles within a triangle add up to 180° and two initially parallel light rays will remain parallel. It should be pointed out that this refers to a universe on the very large scale and, as we have seen, in the region of a massive object, such as a star or galaxy, the space becomes positively curved.

## 9.3 The blueshifts and redshifts observed in the spectra of galaxies

When the spectra of galaxies were first observed in the early 1900s it was found that their observed spectral lines, such as those of hydrogen and calcium, were shifted from the positions of the lines when observed in the laboratory. In the closest galaxies the lines were shifted toward the blue end of the spectrum, but for galaxies beyond our Local Group, the lines were shifted towards the red. This effect is called a redshift or blueshift and the simple explanation attributes this effect to the speed of approach or recession of the galaxy, similar to the falling pitch of a receding train whistle, which we know of as the Doppler effect (Figure 9.3). For speeds which are small compared with the speed of light, then the following simple formula may be used:

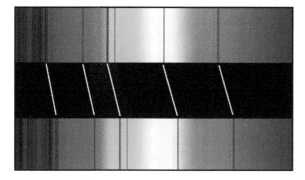

**Figure 9.3** The redshift in the spectral lines from a distant galaxy (below) relative to those observed in our Sun (above). Image: Wikipedia Commons.

$$\Delta f/f = \Delta\lambda/\lambda = v/c$$

where $f$ is the frequency, $\lambda$ is the wavelength, $\Delta f$ and $\Delta\lambda$ are changes in frequency and wavelength, $v$ is the velocity of approach or recession and $c$ is the speed of light.

Some of the earliest observations of redshifts and blueshifts were made by the American astronomer Vesto Slipher. In 1913 he discovered that the Andromeda Galaxy had a blueshift of $300\,km\,s^{-1}$. This implies that the Andromeda and the Milky Way Galaxies are approaching each other due to the gravitational attraction between them but not, as might first appear, at $300\,km\,s^{-1}$. Our Sun is orbiting the centre of our galaxy at about $220\,km\,s^{-1}$ and taking this into account, the actual approach speed is nearer $100\,km\,s^{-1}$.

By 1915 Slipher had measured the shifts for 15 galaxies, 11 of which were redshifted. Two years later, a further six redshifts had been measured and it became obvious that only the nearer galaxies (those within our Local Group) showed blueshifts. From the measured shifts and, using the Doppler formula given above, he was able to calculate the velocities of approach or recession of these galaxies. These data were used by Edwin Hubble in what was perhaps the greatest observational discovery of the last century, and it is perhaps a little unfair that Slipher has not been given more recognition.

## 9.4    The expansion of the universe

In the late 1920s, Edwin Hubble, using the 100 in. Hooker Telescope on Mount Wilson, measured the distances of galaxies in which he could observe a type of very bright variable stars, called Cepheid variables, which vary in brightness with very regular periods. (The method was described in Chapter 8.) He combined these measurements with those of their speed of approach or recession (provided by Slipher) of their host galaxies (measured from the blue or red shifts in their spectral lines) to produce a plot of speed against distance (Figure 9.4). All, except the closest galaxies, were receding from us and he found that the greater the distance, the greater the apparent speed of recession. From this he derived Hubble's Law in which the speed of recession and distance were directly proportional and related by Hubble's constant ($H_0$). The value that is derived from his original data was $\sim500\,km\,s^{-1}\,Mpc^{-1}$. Such a linear relationship is a direct result of observing a universe that is expanding uniformly, so Hubble had shown that we live within an expanding universe. The use of the word 'constant' is perhaps misleading. It would only be a real constant if the universe expanded linearly throughout the whole of its existence. It has not – which is why the subscript is used. $H_0$ is the *current* value of Hubble's constant!

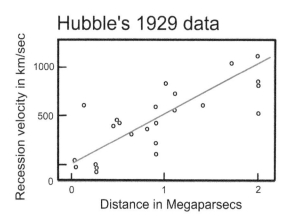

**Figure 9.4** Hubble's plot of recession velocity against distance.

Consider the very simple one-dimensional universe shown in Figure 9.5. Initially the three components are 10 miles apart. Let this universe expand uniformly by a factor of two in 1 h. As seen from the left-hand component, the middle components will have appeared to have moved 10 miles in 1 h whilst the right-hand component will have appeared to move 20 miles – the apparent recession velocity is proportional to the distance.

If one makes the simple assumption that the universe has expanded at a uniform rate throughout its existence, then it is possible to backtrack in time until the universe would have had no size – its origin – and hence estimate the age, known as the Hubble age, of the universe. This is very simply given by $1/H_0$ and, using $500 \, \text{km s}^{-1} \, \text{Mpc}^{-1}$, one derives an age of about 2000 million years:

$$
\begin{aligned}
1/H_0 &= 1 \, \text{Mpc}/500 \, \text{km s}^{-1} \\
&= 3.26 \, \text{million light years}/500 \, \text{km s}^{-1} \\
&= 3.26 \times 10^6 \times 365 \times 24 \times 3600 \times 3 \times 10^5 \, \text{s}/500 \\
&= 3.26 \times 10^6 \times 3 \times 10^5 \, \text{years}/500 \\
&= 1.96 \times 10^9 \, \text{years} \\
&= {\sim}2 \, \text{billion years}
\end{aligned}
$$

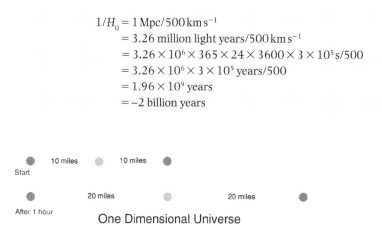

One Dimensional Universe

**Figure 9.5** A simple 'universe' to demonstrate Hubble's Law.

In fact, in all the Friedmann models, the real age must be less than this as the universe would have been expanding faster in the past as you can see in Figure 9.2. In the case of the 'flat' universe the actual age would be two-thirds that of the Hubble age or ~1300 million years old.

### 9.4.1    *A problem with age*

This result obviously became a problem as the age of the Solar System was determined (~4500 million years) and calculations relating to the evolution of stars made by Hoyle and others indicated that some stars must be much older than that, ~10–12 thousand million years old. During the blackouts of World War II, Walter Baade used the 100 in. Hooker Telescope on Mount Wilson to study the stars in the Andromeda Galaxy and discovered that there were, in fact, two types of Cepheid variable. Those observed by Hubble were four times brighter than those that had been used for the distance calibration, and this led to the doubling of the measured galaxy distances. As a result, Hubble's constant reduced to ~250 km s$^{-1}$ Mpc$^{-1}$. There still remained many problems in estimating distances, but gradually the observations have been refined and, as a result, the estimate of Hubble's constant has reduced in value to about 70 km s$^{-1}$ Mpc$^{-1}$.

One of the best determinations is that made by a 'key project' of the Hubble Space Telescope that observed almost 800 Cepheid variable stars in 19 galaxies out to a distance of 108 million light years. Combined with some other measurements the following value was derived:

$$H_0 = 72 +/- 8 \, \text{km s}^{-1} \text{Mpc}^{-1}.$$

Observations of gravitational lenses give a totally independent method of determining the Hubble constant and the best value to date is:

$$H_0 = 71 +/- 6 \, \text{km s}^{-1} \text{Mpc}^{-1}. \text{ (Error is controversial!)}$$

Observations by the WMAP spacecraft (see Sections 9.10 and 9.11) have given another independent value:

$$H_0 = 73.5 +/- 3.5 \, \text{km s}^{-1} \text{Mpc}^{-1}.$$

Combining WMAP data with other cosmological data gives:

$$H_0 = 70.8 +/- 1.6\,\mathrm{km\,s^{-1}\,Mpc^{-1}}.$$

These are all in very good agreement so it is unlikely that the true value of Hubble's constant will differ greatly from $71\,\mathrm{km\,s^{-1}\,Mpc^{-1}}$.

## Exercise: Use the method given above to show that the Hubble age based on the current value of Hubble's constant is ~14 000 million years

A Hubble age of ~14 000 million years corresponds to the age of a 'flat' universe of only ~9300 million years. From observations of globular clusters (which contain some of the oldest stars in the universe) and of the white dwarf remnants of stars, we suspect that the universe is somewhat older than 12 000 million years. Hence, if we believe the current value of Hubble's constant, there is still an age problem with the standard Friedmann Big Bang models.

### The cosmological redshift

It is possible to characterize the redshift (or blueshift) of a galaxy by the relative difference between the observed and emitted wavelengths of an object. This is given the dimensionless quantity, $z$. From the definitions of $z$ it is then possible to derive the alternate forms shown below:

$$z = \frac{\lambda_{observed} - \lambda_{emitted}}{\lambda_{emitted}} \qquad z = \frac{f_{emitted} - f_{observed}}{f_{observed}}$$

$$1 + z = \frac{\lambda_{observed}}{\lambda_{emitted}} \qquad 1 + z = \frac{f_{emitted}}{f_{observed}}$$

In the above, the blueshift and redshift were regarded as being due to the Doppler effect, and this would be perfectly correct when considering the blue shifts shown by the galaxies in the Local Group. However, in the cases of galaxies beyond our Local Group there is a far better way of thinking about the cause of the redshifts that are observed by us. It is not right to think of the galaxies (beyond the movements of those in our Local Group) moving through space but, rather, that they are being carried apart by the expansion of space. A nice analogy is that of baking a currant bun. The dough is packed with currants and then baked. When taken out of the oven the bun will (hopefully) be bigger and thus the currants will be further apart. They will not have moved *through* the dough, but will have been carried apart by the *expansion* of the dough.

As Hubble showed, the universe is expanding so that it would have been smaller in the past. When a photon was emitted in a distant galaxy corresponding to a specific spectral line, the universe would have been smaller. In the time it has taken that photon to reach us whilst the photon has travelled through space, the universe has expanded and this expansion has stretched, by exactly the same ratio, the wavelength of the photon. This increases the wavelength so giving rise to a redshift that we call the 'cosmological redshift'. A simple analogy is that of drawing a sine wave (representing the wavelength of a photon) onto a slightly blown up balloon. If the balloon is then blown up further, the length between the peaks of the sine wave (its wavelength) will increase.

This gives a very nice interpretation of the parameter $z$. If we find that a galaxy is observed at a redshift of $z$, then, by adding 1 to that value we can find the ratio of the wavelengths of the photons observed to those emitted. This simply means that we 'see' the galaxy at a time when the universe was smaller by just this ratio, $(1 + z)$.

For example, if we observe a quasar whose redshift is 6.4 (the highest quasar redshift known at the time of writing), then $1 + z = 7.4$, which means that we see the quasar at a time when the universe was 7.4 times smaller than it is now.

For small redshifts, the apparent velocity of recession is given by the simple formula:

$$v = zc$$

As an example, consider the object 3C273, the first quasar that was discovered. Its measured redshift was 0.158 – a 16% shift in wavelength. This implies that it is moving away from us at a speed given by $zc = 0.158 \times 3 \times 10^5\,\mathrm{km\,s^{-1}} = {\sim}47\,000\,\mathrm{km\ s^{-1}}$. The distance can then be found from Hubble's Law and is given by $d = v/H_0 = 47\,000/72\,\mathrm{Mpc}$. This is ${\sim}650\,\mathrm{Mpc}$ corresponding to ${\sim}2100$ million light years – about 1000 times further away than the Andromeda Galaxy. However, at high redshifts, such as that of the quasar referred to above with a redshift of 6.4, the simple formula $v = cz$ no longer applies and a more complex analysis must be used.

## 9.5    The steady state model of the universe

As a result of the 'age' problem many astronomers did not give much credence to the Big Bang models and in 1948, Herman Bondi, Thomas Gold and Fred Hoyle (who disliked the idea of an instantaneous origin of the universe) proposed an alternative theory called the Steady State theory. All cosmological theories embrace what is called the cosmological principle that is, on the large scale at any given time, the view of the universe from any location within it will be the same. Bondi, Gold and Hoyle extended this principle to give what they called the perfect cosmological principle where the words 'at any given time' were replaced by 'for all time'. Their universe was unchanging on the large scale. That did not mean that it was not expanding. At the heart of their theory was the idea that, as the galaxies moved further apart due to the expansion of the universe, new matter, in the form of hydrogen, was created in the space between them which eventually formed new galaxies to keep the observed density of galaxies constant. The universe had no beginning and will have no end and is, as the theory's name implies, in a 'steady state'. As new matter is continuously being created, it is also called the theory of continuous creation.

## 9.6    Big Bang or Steady State?

In the early 1960s observational tests were made to decide between the two theories. Suppose one could measure the galaxy density close to us – to give the number of galaxies per cubic megaparsec. As these galaxies are close to us we see them essentially at the present time. If we could then measure the density of galaxies in the far universe, we would be measuring it at some time in the past. In the Steady State model these results should be the same, but in the Big Bang model the density should have been higher in the past. Martin Ryle at Cambridge attempted such measurements by counting radio sources. Though there were problems with his initial data, these results did finally indicate a greater density of radio sources in the past so disproving the Steady State theory. The deathblow to the Steady State theory came in 1963 when radiation, believed to have come from the Big Bang, was discovered. The origin, discovery and study of this radiation will form a major part of the cosmological story to follow.

## 9.7    The cosmic microwave background

It was the American physicist, George Gamow, who first realised that the Big Bang should have resulted in radiation that would still pervade the universe. This radiation is now called the cosmic microwave background (CMB). Initially in the form of very high-energy gamma rays, the radiation became less energetic as the universe expanded and cooled, so that by a time some 300–400 thousand years after the origin the peak of the radiation was in the optical part of the spectrum.

Up to that time the typical photon energy was sufficiently high to prevent the formation of hydrogen and helium atoms and thus the universe was composed of hydrogen and helium nuclei and free electrons – so forming a plasma. The electrons would have scattered photons rather as water droplets scatter light in a fog and thus the universe would have been opaque. This close interaction between the matter and radiation in the universe gave rise to two critical consequences: first, the radiation would have a black body spectrum corresponding to the then temperature of the universe; and secondly, the distribution of the nuclei and electrons (normal matter) would have a uniform density except on the very largest scales.

We will return to the second consequence later, but now will continue with the first. As the universe expanded and cooled there finally came a time, ~380 000 years after the origin, when the typical photon energy became low enough to allow atoms to form. There were then no free electrons left to scatter radiation so the universe became transparent. This is thus as far back in time as we are able to see. At this time the universe had a temperature of ~3000K. Since that time, the universe has expanded by about 1000 times. The wavelengths of the photons that made up the CMB will also have increased by 1000 times and so will now be in the far infrared and radio part of the spectrum but would still have a black body spectrum. The effective black body temperature of this radiation will have fallen by just the same factor and would thus now be ~3 K.

### 9.7.1    *The discovery of the cosmic microwave background*

Radio astronomers Arno Penzias and Robert Wilson serendipitously discovered this background radiation in 1963. However, incontrovertible proof as to its origin had to wait until 1992 when the COBE satellite was able to show that the background radiation had the precise black body spectrum that would have been expected (Figure 9.6). It is worth telling a little of the Nobel Prize winning story of its discovery. Penzias and Wilson had been given use of the telescope and receiver that had been used for the very first passive satellite communication experiments using a large aluminium covered balloon called 'Echo'. It had been designed to minimize any extraneous noise that might enter the horn shaped telescope and the receiver was one of the best in the world at that time.

They tested the system thoroughly and found that there was more background noise produced than they expected. They wondered if it might have been caused by pigeons nesting within the horn – the pigeons, being at ~290 K, would radiate radio noise – and bought a pigeon trap (now in the Smithsonian Air and Space Museum in Washington) to catch the pigeons. They took the pigeons many miles away and released them but, as pigeons do, they returned

**Figure 9.6** Arno Penzias and Robert Wilson and the Holmdel antenna. Image: Wikipedia Commons.

and had to be removed by a local pigeon expert. During their time within the horn antenna, the pigeons had covered much of the interior with what, in their letter to the journal *Nature*, was called 'a white dielectric substance' – we might call it 'guano'. This was cleaned out, but having removed both the pigeons and the guano there was no substantial difference. The excess noise remained the same wherever they pointed the telescope – it came equally from all parts of the sky.

An astronomer, Bernie Burke, when told of the problem suggested that they contact Robert Dicke at Princeton University. Dicke had independently theorized that the universe should be filled with radiation resulting from the Big Bang and was building a horn antenna on top of the Physics department in order to detect it. Learning of Penzias and Wilson's observations, Dicke immediately realised that his group had been 'scooped' and told them that the excess noise was not caused within their horn antenna or receiver but that their observations agreed exactly with the predictions that the universe would be filled with radiation left over from the Big Bang. Dicke was soon able to confirm their result, and it was perhaps a little unfair that he did not share in the Nobel Prize. The average temperature of the CMB is 2.725 K. Thirty years later COBE's measurements were able to show that the CMB had the precise black body spectrum that would result from the Big Bang scenario. Since then, it has been very difficult to refute the fact that there was a Big Bang.

## 9.8    Inflation

By the 1970s, problems with the standard Big Bang models had arisen. Observations had shown that the universe was very close to being 'flat', $\Omega \sim 1$, and the Big Bang theory gives no particular reason why this should be so. Any curvature that the universe has close to its origin tends to get enhanced as the universe ages – a slightly positively curved space becomes more and more so and vice versa. In fact, had not $\Omega$ been in the range 0.999999999999999–1.000000000000001 just 1 s after its origin the universe could not be as it is now! This is incredibly fine tuning, and there is nothing in the standard Big Bang theory to explain why this should be so. This is called the flatness problem.

A second problem is known as the horizon problem. The universe appears to have exactly the same properties – specifically the observed temperature of the CMB – in opposing directions. The CMB from one direction has taken nearly 14 000 million years to reach us, and the same from the opposing direction. In the standard Big Bang models there has not been sufficient time to allow radiation to travel from one of these regions to the other – they cannot 'know' what each other's temperature is, as this information cannot travel faster than the speed of light. So why are they at precisely the same temperature?

These problems were addressed with the idea of inflation, first proposed by Alan Guth and refined by others. In this scenario the whole of the visible universe would have initially been contained in a volume of order the size of a proton. Some $10^{-35}$ s after the origin this volume of space began to expand exponentially and increased in size by a factor of order $10^{26}$ in a time of $\sim 10^{-32}$ s to the size of a sphere a metre or more in size (Figure 9.7). This massive expansion of space would force the geometry of space to become 'flat', just as the surface of a balloon appears to become flatter and flatter as it expands. (Hence one would naturally get a 'flat' universe.)

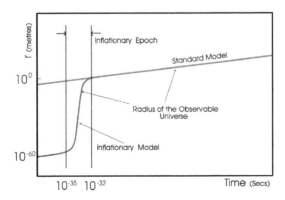

**Figure 9.7** The inflationary epoch in the early universe.

Inflation would also ensure that the whole of the visible universe would have uniform properties so also addressing the horizon problem. This is a result of the fact that, prior to the inflationary period, the volume of space that now forms the visible universe was sufficiently small that radiation could easily travel across it and so give it a uniform temperature.

## 9.9    The Big Bang and the formation of the primeval elements

Half of the gravitational potential energy that arose from this inflationary period was converted into kinetic energy from which arose an almost identical number of particles and antiparticles, but with a very small excess of matter particles (~1 part in several billion). All of the antiparticles annihilated with their respective particles leaving a relatively small number of particles in a bath of radiation. The bulk of this 'baryonic matter' was in the form of quarks that, at about 1 s after the origin, grouped into threes to form protons and neutrons. (Two up quarks and one down quark form a proton, and one up quark and two down quarks a neutron. The up quark has $+2/3$ charge and the down quark $-1/3$ charge so the proton has a charge of $+1$ and the neutron 0 charge.)

An almost equal number of protons and neutrons were produced but, as free neutrons are unstable with a half-life of 10.6 min, the only ones to remain were those that were incorporated into helium nuclei comprising two protons and two neutrons. So, after a few minutes, the normal (baryonic) matter in the universe was very largely composed of hydrogen nuclei (protons), helium nuclei (alpha particles) and electrons – with one electron for each proton. (We now believe that several times more 'dark matter', whose constitution will be discussed below, was also created.)

## 9.10    The 'ripples' in the Cosmic Microwave Background

Observations by the COBE spacecraft first showed that the CMB did not have a totally uniform temperature and, since then, observations from the WMAP spacecraft, balloons and high mountain tops have been able to make maps of these so called 'ripples' in the CMB – temperature fluctuations in the observed temperature of typically 60 μK (Figure 9.8).

Why are these small variations present? To answer this we need to understand a little about dark matter. Though not yet directly detected, its presence has been inferred from a wide variety of observations which will be discussed in detail later.

As described in Section 9.7 about the CMB, for ~380 000 years following the Big Bang, the matter and radiation were interacting as the energy of the photons was sufficient to ionize the atoms giving rise to a plasma of nuclei and free electrons. This gives rise to two results:

**Figure 9.8** All-sky map of the CMB ripples produced by 5 years of WMAP data in March 2008. Image: NASA/WMAP Science team.

(1)    The radiation and matter were in thermal equilibrium and the radiation will thus have had a black body spectrum.

(2)    The plasma of nuclei and electrons will have been very homogeneous as the photons acted rather like a whisk beating up a mix of ingredients.

It is the second of these that is important to the argument that follows. When the temperature drops to the point that atoms can form, the matter can begin to clump under gravity to form stars and galaxies. Simulations have shown that, as the initial gas is so uniformly distributed, it would take perhaps 8–10 billion years for regions of the gas to become sufficiently dense for this to happen. However, we know that galaxies came into existence around 1 billion years after the Big Bang. Something must have aided the process. We believe that this was dark matter that was not composed of normal matter and which is called non-baryonic dark matter. As this would not have been coupled to the radiation, it could have begun to gravitationally 'clump' immediately after the Big Bang. Thus, when the normal matter became decoupled from the photons, there were 'gravitational wells' in place formed by concentrations of dark matter. The normal matter could then quickly fall into these wells, rapidly increasing its density and thus greatly accelerating the process of galaxy formation.

## 9.11    How dark matter affects the cosmic microwave background

The concentrations of dark matter that existed at the time the CMB originated have an observable effect due to the fact that if radiation has to 'climb out' of a gravitational potential well it will suffer a type of red shift called the 'gravitational red shift'. Hence, the photons of the CMB that left regions where the dark matter had clumped would have had longer wavelengths than those that left regions with less dark matter. This causes the effective black body temperature of photons coming from denser regions of dark matter to be less than those from sparser regions – thus giving rise to the temperature fluctuations that are observed. As such observations can directly tell us about the universe as it was just 380 000 or so years after its origin it is not surprising that they are so valuable to cosmologists!

This is no easy matter. The CMB needs to be observed at millimetre radio wavelengths that are masked by emission from water vapour in the Earth's atmosphere. Consequently, experiments have been flown in satellites (COBE and WMAP), balloons (Boomerang and Maxima) or located at high dry sites on Earth such as the Atacama Desert in Chile at a height of 16 000 ft (5000 m) or on the flanks of Mount Teide in Tenerife (the CBI and VSA experiments, respectively) (Figure 9.9). Another very good site, where the DASI experiment is located, is at the South Pole where it is so cold that the water vapour is largely frozen out of the atmosphere!

Observations of the CMB also enable us to measure the curvature of space. The photons that make up the CMB have travelled across space for billions of years and will thus have been affected by the curvature of space. We can calculate the spatial form of the fluctuations at the time of their origin and it is possible to simulate the expected pattern of fluctuations if space were negatively curved, positively curved or flat (Figure 9.10) so that, if astronomers could map these fluctuations accurately, then it would be possible to measure the curvature of space.

Consider this analogy. Imagine looking at a distant wall covered with a repeating pattern of wallpaper. Looking directly at the wall we could measure the observed angular spacing of the items that make up the pattern. This is analogous to observing through flat space. Suppose now we observed through a

**Figure 9.9** The cosmic background imager in the Atacama Desert, Chile. Image: CBI/Caltech/NSF.

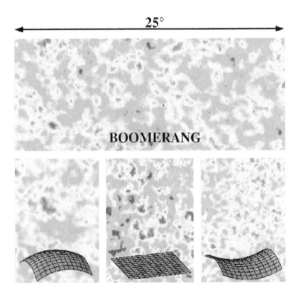

**Figure 9.10** Boomerang observed fluctuations in the CMB that are consistent with space being 'flat'. –(Above) Boomerang map. (Below) What would be observed with positively curved, flat and negatively curved space. Image: The international BOOMERANG consortium.

concave lens. The pattern would appear smaller and the angular separation less; analogous to observing through negatively curved space. If we then observed through a convex lens (as used in a magnifying glass) the angular separation would increase; analogous to observing through positively curved space. Thus, by comparing the observed fluctuation pattern with that impressed on the CMB at its origin allows us to measure the curvature of space. The results of these observations confirm, without exception, that space is flat to within 1–2%: $\Omega = 1$.

## 9.12    The hidden universe: dark matter and dark energy

If, as inflation predicts and observations confirm, space has zero curvature, it is possible to calculate the average density and the total mass/energy content of the visible universe. If the total mass was $M$, then it appears that the best estimate of the mass of the visible matter, stars and excited gas, is about $0.01M$. Hence, 99% of the content of the universe is invisible! The first question to ask is whether this invisible content is normal (baryonic) matter that just does not emit light such as gas, dust, or objects such as brown dwarfs, neutron stars or black holes. These latter objects are called MACHOs (Massive Astronomical Compact Halo Objects) as many would reside in the galactic halos that extend around galaxies.

There are two pieces of evidence that indicate that the total amount of normal matter in the universe is only ~4% of the total mass/energy content. The first depends on measurements of the relative percentages of hydrogen, helium and lithium and their isotopes that were formed in the Big Bang. These are very sensitive to the baryon to photon ratio and put an upper limit of baryonic matter at about 4%. The second line of evidence is that if a significant amount of mass was in the form of MACHOs then gravitational microlensing studies (as have discovered a number of planets) would have detected them. Though we know that, for example, pulsars are found in the galactic halo the total mass of these and other MACHOs cannot explain the missing matter. So we still have to account for ~96% of the total mass/energy content of the universe! From several lines of observational evidence it is believed that a substantial part of this is in the form of non-baryonic dark matter – usually just called dark matter.

## 9.12.1     *Evidence for dark matter*

In Chapter 8 it has been described how one of the first pieces of evidence for the presence of dark matter emerged – dark matter being needed to explain the anomalous rotation curves of galaxies. In this section further evidence, some very recent, will be discussed.

### Cluster dynamics

The first evidence of a large amount of unseen matter came from observations made by Fritz Zwicky in the 1930s. He studied the Coma cluster of galaxies, 321 million light years distant, and observed that the outer members of the cluster were moving at far higher speeds than were expected. Suppose a cluster of galaxies was created which was not in motion. Gravity would cause it to collapse down into a single giant body. If, on the other hand, the galaxies were initially given very high speeds relative one to another, their kinetic energy would enable them to disperse into the universe and the cluster would disperse, just as a rocket travelling at a sufficiently high speed could escape the gravitational field of the Earth. The fact that we observe a cluster of galaxies many billions of years after it was created implies that there must be an equilibrium balancing the gravitational pull of the cluster's total mass and the average kinetic energy of its members. This concept is enshrined in what is called the virial theorem so that, if the speeds of the cluster members can be found, it is possible to estimate the total mass of the cluster. Zwicky carried out these calculations and showed that the Coma cluster must contain significantly more mass than could be accounted for by its visible content.

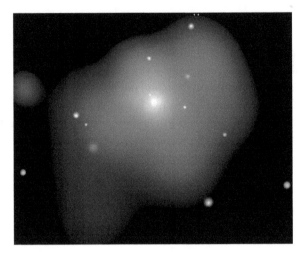

**Figure 9.11** NASA'a Chandra X-ray satellite image of hot gas surrounding the galaxy NGC 4555. Image: NASA/CXC/E.O'Sullivan et al.

### Gas entrapment

NASA's Chandra X-ray Observatory has revealed that the elliptical galaxy NGC 4555 is embedded in a cloud of gas having a diameter of about 400 000 light years and a temperature of 10 million°C (Figure 9.11). At this temperature the molecules of gas would be travelling at very high speeds and the mass of the stars within the galaxy would be far too low to prevent its escape. For the gas to remain in the vicinity of the galaxy the total mass of the system must be about 10 times the combined mass of the stars in the galaxy and about 300 times that of the gas cloud.

### Gravitational lensing

Earlier, the formation of multiple images of a distant object by a foreground galaxy has been discussed. On a much larger scale the mass of a cluster of galaxies can distort the images of more distant objects. The image of the Abell 2218 cluster is a wonderful example showing images of more distant galaxies that have been distorted into arcs (Figure 9.12). The amount of distortion will be a function of the total mass of the intervening cluster, so this gives a way of estimating the total mass of galaxy clusters, confirming the existence of dark matter. Using this technique, astronomers have even shown how the distribution of dark matter has become more 'clumpy' over the last 6 billion years.

**Figure 9.12** The Abell 2218 cluster imaged by the Hubble Space Telescope. Image: NASA, A. Fruchter and the ERO team, STScI.

### 9.12.2     *How much non-baryonic dark matter is there?*

There are several ways of estimating the amount of dark matter. One of the most direct is based on the detailed analysis of the fluctuations in the CMB. The percentage of dark matter has an observable effect, and the best fit to current observations corresponds to dark matter making up ~23% of the total mass/energy content of the universe. Other observations support this result. This then leaves two further questions: what is dark matter and what provides the remaining 73% of the total mass/energy content?

### 9.12.3     *What is dark matter?*

The honest answer is that we do not really know. The standard model of particle physics does not predict its existence and so extensions to the standard theory (which have yet to be proven) have to be used to predict what it might be and suggest how it might be detected.

Dark matter can be split into two possible components: hot dark matter would be made up of very light particles moving close to the speed of light (hence hot) whilst cold dark matter would comprise relatively massive particles moving more slowly. Simulations that try to model the evolution of structure in the universe – the distribution of the clusters and superclusters of galaxies – require that most of the dark matter is 'cold' but astronomers do believe that there is a small component of hot dark matter in the form of neutrinos. There are vast numbers of neutrinos in the universe but they were long thought to have no mass. However, recent observations attempting to solve the solar neutrino problem discussed earlier, show that neutrinos can oscillate between three types (electron, tau and muon). This implies that they must have some mass but current estimates put this at less than one-millionth of the mass of the electron. As a result they would only make a small contribution to the total amount of dark matter – agreeing with the simulations.

A further confirmation of the fact that hot dark matter is not dominant is that, if it were, the small scale fluctuations that we see in the WMAP data would have been 'smoothed' out and the observed CMB structure would have shown far less detail.

### Axions

One possible candidate for cold dark matter is a light neutral axion whose existence was predicted by the Peccei–Quinn theory in 1977. There would be of order 10 trillion in every cubic centimetre! If axions exist they could theoretically change into photons (and vice versa) in the presence of a strong magnetic field. One possible test would be to attempt to pass light through a wall. A beam of light is passed through a magnetic field cavity adjacent to a light barrier. A photon might rarely convert into an axion which could easily pass through the wall where it would pass through a second cavity where (again with an incredibly low probability) it might convert back into a photon!

Another experiment at the Lawrence Livermore Laboratory is searching for microwave photons within a tuned cavity that might result from an axion decay, whilst in Italy polarized light is being passed back and forth millions of times through a 5 T field. If axions exist, photons could interact with the field and become axions; causing a very small anomalous rotation of the plane of polarization. The most recent results do indicate the existence of axions with a mass of ~3 times that of the electron, but this has to be confirmed and there may well be other causes for the observed effect on the light.

### WIMPS

An extension to the standard model of particle physics, called 'super symmetry', suggests that WIMPS (Weakly Interacting Massive Particles) might be a major constituent of cold dark matter. A leading candidate is the neutralino – the lightest neutral super symmetric particle. Billions of WIMPS could be passing through us each second! Very occasionally they will interact with the nucleus of an atom making it recoil – rather like the impact of a moving billiard ball with a stationary one. In principle, but with very great difficulty, these interactions can be detected.

Some possible ways of detecting the nuclear recoils resulting from a WIMP interaction:

(1)   In semiconductors such as silicon and germanium, electric charge is released by the recoiling atom. This ionisation can be detected and measured.

(2)   In certain types of crystals and liquids that are termed 'scintillators', flashes of light are emitted as the atom slows down. The light, whose amount depends on the recoil energy, can be detected by photomultiplier tubes.

(3)   In crystals, the recoil energy is transformed into vibrations called phonons. At room temperature these would be lost amongst thermally induced vibrations within the crystal but, if the crystal is cooled to close to absolute zero, they could be detected.

Though a million WIMPS might pass through every square centimetre each second, they will very rarely interact with a nucleus. It is estimated that within a 10 kg detector only one interaction might occur, on average, each day. To make matters worse, we are being bombarded with cosmic rays which, being made of normal matter, interact very easily. Any WIMP interactions would be totally swamped! One way to greatly reduce the number of cosmic rays entering a detector is to locate it deep underground – such as at the bottom of the Boulby Potash Mine in north Yorkshire at a depth of 1100 m. At this depth, the rock layers will have stopped all but one in a million cosmic rays. In contrast, only about three in a billion WIMPs would have interacted with nuclei in the rock above the detector.

In addition, natural radioactivity in the rocks surrounding the experimental apparatus increases the 'noise' which can mask the WIMP interactions, so the detectors are surrounded by radiation shields of high purity lead, copper wax or polythene and may be immersed within a tank of water. The chosen detectors may also emit alpha or beta particles so care must be taken over the materials from which they are made. Photomultiplier tubes (to detect scintillation) cause a particular problem and 'light guides' are used to transfer the light from the crystal, such as sodium iodide, in which the interaction takes place to the shielded photomultiplier tube.

## Possible success?

One possible way to show the presence of WIMP interactions in the presence of those caused by local radioactivity is due to the fact that, in June, the motion of the Earth around the Sun ($29.6 \text{ km s}^{-1}$) is in the same direction as that of the Sun in its orbit around the galactic centre ($232 \text{ km s}^{-1}$). Hence, the Earth would sweep up more WIMPS than in December when the motions are opposed. The difference is ~7%, so one might expect to detect more WIMPs in June than in December. As the number of interactions from local radioactivity should remain constant, this gives a possible means of making detection. In the DArk MAtter (DAMA) experiment at the Gran Sasso National Laboratory, 1400 m underground in Italy, observations have been made of scintillations within 100 kg of pure sodium iodide crystals. The results of seven annual cycles have given what is regarded as a possible detection but, again, there may be other explanations.

## 9.12.4    *Dark energy*

Normal and dark matter can between them account for some 27% of the total mass/energy of the universe. It appears that the majority, some 73%, must be something else. It is thought to be a form of energy latent within space itself that is totally uniform throughout space. In fact this could be exactly what was invoked by Einstein to make his 'static' universe – the cosmological constant or lambda ($\Lambda$) term. A positive $\Lambda$ term can be interpreted as a fixed positive energy density that pervades all space and is unchanging with time. Its net effect would be repulsive. There are, however, other options and a range of other models are being explored where the energy is time dependent. These are given names such as 'quintessence', meaning 5th force. As the total amount of this energy and its repulsive effects are proportional to the volume of space, the effects of dark energy should become more obvious as the universe ages and its size increases.

In all of the Friedmann models of the universe, the initial expansion slows with time as gravity reins back the expansion, and the expansion rate would never increase. However, if there is a component in the universe whose effect is repulsive and increasing with the volume of space, the scale size of the universe will vary in a quite different way with time. Initially, when the volume of the universe is small, gravity will dominate over dark energy and the initial expansion rate of the universe will slow – just as in the Friedmann models – but there will come a point when the repulsive effects of the dark energy will equal and then overcome gravity and the universe will begin to expand at an ever increasing rate. If this is the case, distant galaxies will be further away from us than would have been the case in the Friedmann models.

## 9.12.5    *Evidence for dark energy*

In the 1990s it became possible to measure the distance to very distant galaxies. We can estimate the distance to remote objects if we have what is called a standard candle – an object of known brightness some of which have been observed at known distances nearby. Hubble used such a technique to measure the distances of galaxies using Cepheid variable stars of known peak brightness. These had first been observed in the Small Magellanic Cloud (SMC) at a known distance from us. Suppose one of these stars is observed in a distant galaxy and appears 1/10000th as bright as a similar Cepheid in the SMC. Assuming no extinction by dust it would, from the inverse square law, be at 100 times the distance of the SMC.

However, though Cepheid variable stars are some of the brightest stars known, there is a limit to distances that can be measured by using them. Something brighter is required. For a short time, supernovae are the brightest objects in the universe and there is one variant, called a Type 1a supernova, that is believed to have well calibrated peak brightness. It might be useful to consider an analogy. Imagine a ball of plutonium of less than critical mass. If one then gradually added additional plutonium uniformly onto its surface it would, at some point in time, exceed the critical mass and explode. The power of this explosion should be the same each time the experiment is carried out as a sphere of plutonium has a well defined critical mass. As you will see, this is rather similar to what occurs when a Type 1a supernova occurs.

A Type 1a supernova occurs in a binary system. The more massive star of the pair will evolve to its final state first and its core may become a white dwarf about the size of the Earth. Later its companion will become a red giant and its size will dramatically increase. Its outer layers may then be attracted onto the surface of the white dwarf whose mass will thus increase (Figure 9.13). At some critical point, when its mass nears the Chandrasekhar Limit of roughly 1.44 times the mass of the Sun, the outer layers will ignite and in the resulting thermonuclear explosion the entire white dwarf star will be consumed. As all such supernovae will explode when they reach the same total mass, it is expected that they will all have similar peak brightness (about 5 billion times brighter than the Sun) and should thus make excellent standard candles.

As Type Ia supernovae are so bright, it is possible to see them at very large distances. The brightest Cepheid variable stars can be seen at distances out to about 10–20 Mpc (~32–64 million light years). Type Ia supernovae are approximately

**Figure 9.13** Material accreting onto a white dwarf. Image: STScI, NASA, Wikipedia Commons.

14 magnitudes brighter than Cepheid variables, and are thus about a quarter of a million times brighter. They can thus be can be seen about 500 times further away corresponding to a distance of around 1000 Mpc – a significant fraction of the radius of the known universe. However, supernovae are rare with perhaps one each 300 years in a typical spiral galaxy. Observations are now being made of thousands of distant galaxies on a regular basis and sophisticated computer programs look for supernovae events. Once initially detected, observations continue to look for the characteristic light curve of a Type 1a supernova which results from the radioactive decay of nickel-56; first to cobalt-56 and then to iron-56.

Hubble (and later others) plotted the apparent expansion velocity of galaxies against their distance and produced a linear plot. This plot would not be expected to continue linearly out to very great distances due to changes in the expansion rate of the universe over time. For the critical (or zero curvature) universe which the CMB observations imply, the curve would have been expected to fall below the linear line for great distances. Observations of distant Type Ia supernovae have recently enabled far greater distances to be measured which, together with the corresponding redshifts, have enabled the Hubble plot to be extended to the point where the plot would no longer be linear. As expected, the plot is no longer linear but, to great surprise, the curve falls above the linear extrapolation, not below. This implies that the expansion of the universe is speeding up – not slowing down as expected – and is thus evidence that dark energy exists.

### 9.12.6    *The nature of dark energy*

Dark energy is known to be very homogeneous, not very dense (about $10^{-29}\,\mathrm{g\,cm^{-3}}$) and appears not to interact through any of the fundamental forces other than gravity. This makes it very hard to detect in the laboratory.

The simplest explanation for dark energy is that a volume of space has some intrinsic, fundamental energy as hypothesized by Einstein with his cosmological constant. Einstein's special theory of relativity relates energy and mass by the relation $E = mc^2$ and so this energy will have a gravitational effect. It is often called a vacuum energy because it is the energy density of empty vacuum. In fact, most theories of particle physics predict vacuum fluctuations that would give the vacuum exactly this sort of energy. One can perhaps get some feeling of why a pure vacuum can contain energy by realizing that it is not actually empty! Heisenberg's Uncertainty Principle allows particles to continuously come into existence and (quickly) go out of existence again. A pure vacuum is seething with these virtual particles!

According then to Heisenberg's Uncertainty Principle there is an uncertainty in the amount of energy that can exist. This small uncertainty allows non-zero energy to exist for short intervals of time defined by:

$$\Delta E \times \Delta T \text{ is of order } h/2\pi$$
where $h$ is Planck's constant $(6.626 \times 10^{-34}\,\text{m}^2\,\text{kg}\,\text{s}^{-1})$.

As a result of the equivalence between matter and energy, these small energy fluctuations can produce particles of matter (a particle and its antiparticle must be produced simultaneously) which come into existence for a short time and then disappear. As an example, consider a proton and the antiproton which have masses of $1.7 \times 10^{-24}\,\text{g}$. If a virtual pair were to be created, their equivalent energy would be (from $E = mc^2$) $3 \times 10^{-3}\,\text{erg}$ and thus they could only exist for a time of order $3 \times 10^{-25}\,\text{s}$.

A number of experiments have been able to detect this vacuum energy. One of these is the Casimir experiment in which, in principle, two metal plates are placed very close together in a vacuum. In practice it is easier to use one plate and one plate which is part of a sphere of very large radius. One way to think of this is that the virtual particles have associated wavelengths – the wave particle duality. Virtual particles whose wavelengths are longer than the separation of the plates cannot exist between them so there are more virtual particles on their outer sides and this imbalance gives the effect of an attractive force between the plates.

An interesting analogy is when two ships sail alongside each other to transfer stores or fuel in the open sea in conditions with little wind but a significant swell. Between them, only waves whose wavelength is smaller than the separation of the hulls can exist but on the outside all wavelengths can be present. This inequality gives rise to a force that tends to push the two ships apart, thus requiring that the ships actively steer away from each other.

The cosmological constant is the simplest solution to the problem of cosmic acceleration with just one number successfully explaining a variety of observations and has become an essential feature in the current standard model of cosmology. It is called the lambda–cold dark matter model, as it incorporates both cold dark matter and the cosmological constant, and can be used to predict the future of the universe.

## 9.13     The makeup of the universe

The observations of the CMB, Type Ia supernovae, Hubble's constant and the distribution of galaxies in space all now give a consistent model of the universe. It appears that normal matter accounts for just ~4%, dark matter ~23% with the remaining ~73% of the total mass energy content of the universe being in the form of dark energy. Over the next few years, as the CMB observations are refined, we will have pretty accurate values for these percentages.

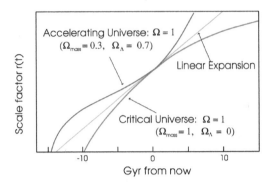

**Figure 9.14** The scale size of the universe with time.

Figure 9.14 shows how we believe that the scale size of the universe has changed with time in the past and how it will expand ever faster in the future. You will see that the actual age of the universe is similar to the Hubble age with a value of about 13–14 thousand million years. [The latest values from the Jodrell Bank's 'Very Small Array' (VSA) and WMAP are 13.6 and 13.7 thousand million years, respectively.]

## 9.14    A universe fit for intelligent life

The very fact that you are reading this book tells us that our universe has just the right properties for intelligent life to have evolved. But why should this be so? As eloquently described in a book *Just Six Numbers* by Martin Rees, there are a number of parameters that have a major influence on how universes can evolve and how stars produce elements that are needed for life. Two of these have been already been covered in this chapter; the constants omega, $\Omega$, and lambda, $\Lambda$. If $\Omega$ had been higher the universe would have rapidly collapsed without allowing life a chance to evolve; if it had been smaller galaxies and stars would not have formed. In addition, if $\Lambda$, which is surprisingly small, had been larger it would have prevented stars and galaxies forming.

You have also read in this chapter how the galaxies formed as a result of fluctuations in the density of the primeval universe – the so-called 'ripples' that are observed in the CMB. The parameter that defines the amplitude of the ripples has a value of $\sim 10^{-5}$. If this parameter were smaller the condensations of dark matter that took place soon after the Big Bang (and were crucial to the formation of the galaxies) would have been both smaller and more spread out resulting in rather diffuse galaxy structures in which star formation would be very inefficient and planetary systems could not have formed. If the parameter had been less than $10^{-6}$, galaxies would not have formed at all! However, if this parameter were greater than $10^{-5}$ the scale of the 'ripples' would be greater and giant structures, far greater in scale than galaxies, would form and then collapse into super-massive black holes – a violent universe with no place for life!

One parameter of our universe is so well known that it is barely given a moment's thought – the number of spatial dimensions, three. If this were either two or four, life could not exist.

Though we perceive gravity to be a 'strong' force (because we are close to a very massive body) it is actually incredibly weak in comparison with the electrostatic forces that control atomic structures and, for example, cause protons to repel each other. The factor is of order $\sim 10^{-36}$. Let us suppose gravity was stronger by a factor of a million. On the small scale, that of atoms and molecules, there would be no difference, but it would be vastly easier to make a gravitationally bound object such as the Sun and planets but whose sizes would be about a billion times smaller. Any galaxies formed in the universe would be very small with tightly packed stars whose interactions would prevent the formation of stable planetary orbits. The tiny stars would burn up their fuel rapidly allowing no time for life to evolve even if there were suitable places for it to arise. Our intelligent life could not have arisen here on Earth if this ratio had been even slightly smaller than its observed value.

Einstein's famous equation, $E = mc^2$, relates the amount of energy that can be extracted from a given amount of mass, so the value of $c$ is obviously fundamentally important. In practice only a small part of the energy bound up in matter can be released, as in the conversion of hydrogen to helium. This process releases 0.7% of the mass of the four protons that form helium – a percentage closely linked to the strength of the strong nuclear force. This parameter, 0.007, has been called 'nuclear efficiency'. However, if this value was too small, say 0.006, the sequence of reactions that build up helium could not take place. In the first of these reactions, two protons form a deuterium nucleus but, given a value of 0.006 for the nuclear efficiency, deuterium would be unstable so preventing the further reactions that give rise to helium – stars would be inert. However, if this parameter was 0.008, meaning that nuclear forces were stronger relative to electrostatic forces, the electrostatic repulsion of two protons would be overcome and they could bind together so no hydrogen would have remained to fuel the stars. A critical reaction in the evolution of stars is the formation of carbon in the triple alpha process. As described earlier, Fred Hoyle played a key role in the understanding of this reaction and pointed out that even a change of a few per cent from the observed value of 0.007 would have severe consequences on the amount of carbon that would be formed in stars – with obvious consequences for life as we understand it.

### 9.14.1    *A 'multiverse'*

So how can it be that all the parameters described above are finely tuned so that we can exist? There are two possible reasons. The first is that our universe was 'designed' by its creator specifically so that it could contain intelligent beings, a view taken by some scientist-theologians. A second view is that there are many universes, each with different properties; the term 'multiverse' has been applied to this view. We have no knowledge of what lies in the cosmos beyond the horizon of our visible universe. Different regions could have different properties; these regions could be regarded as different universes within the overall cosmos. Our part of the cosmos is, like baby bear's porridge, just right.

### 9.14.2    *String theory: another approach to a multiverse*

Theoretical physicists have a fundamental problem. Einstein's General Theory of Relativity that relates to gravity is a classical theory, whereas the other forces are described by quantum mechanics. A 'theory of everything' has yet to be found that can bring together all the fundamental forces. One approach that is being actively pursued is that of string theory. The early string theories envisioned a universe of 10 dimensions, not four, making up a 10-dimensional space–time. The additional six beyond our three of space and one of time are compacted down into tiny regions of space of order $10^{-35}$ m in size and called strings. These are the fundamental building blocks of matter. Different 'particles' and their properties, depend on the way these strings are vibrating – rather like the way a string of a violin can be excited into different modes of vibration to give harmonically related sounds. As these strings move, they warp the space–time surrounding them in precisely the way predicted by general relativity. So string theory unifies the quantum theory of particles and general relativity.

In recent years five string theories have been developed each with differing properties. In one there can be open strings (a strand with two ends) as well as closed strings where the ends meet to form a ring. The remaining four only have closed rings. More recently Ed Witten and Paul Townsend have produced an 11-dimensional 'M-theory' which brings together the five competing string theories into a coherent whole. This 11th dimension (and it not impossible that there could be more) gives a further way of thinking about a multiverse.

We can think of a simple analogy: take a sliced loaf and separate each slice by, say, 1 cm. On each of these slices add some ants. The ants could survive, at least for a while, eating the bread of what is, to them, effectively a two-dimensional universe. They would not be aware of the existence of other colonies of ants on adjacent slices. However, *we* can see that all of these exist within a cosmos that actually has a third dimension.

In just the same way, rather than being individual regions of one large spatially linked cosmos, it could be that other 'universes' exist in their own space–time – hidden from ours within a further dimension.

## 9.15     Intelligent life in the universe

Our universe is suitable for life. How widespread will it be? We suspect that the vast majority of life forms will have the same basic chemistry as our carbon based life. The elements that play a major role in our chemistry (carbon, oxygen and nitrogen) are those first produced in stars and are thus very common. In addition, carbon has the most diverse chemistry of any element. It is not impossible that other life forms could exist based on phosphorus, arsenic and methane, but these would, the author suspects, be far less common.

So, in most cases, we need locations where water is a liquid, such as the surface of a planet in its star's habitable zone or perhaps an under-ice ocean of a satellite warmed by the tidal heating of a nearby giant planet. If we hope that other advanced civilizations such as our own exist then significant periods of time are needed – to allow the simple life forms that may arise a chance to evolve.

In 1960, Frank Drake, who the previous year had made the first search for signals from an extra-terrestrial intelligence in Project Ozma, gathered a group of eminent scientists to try to estimate how likely it was that other intelligent civilizations existed in the Galaxy and might perhaps be transmitting signals that we could detect by observing programs covered by the term SETI (Search for ExtraTerrestrial Intelligence).

### 9.15.1     *The Drake equation*

This group produced what has become known as the Drake equation, which has two parts. The first part attempts to calculate how often intelligent civilizations arise in the galaxy and the second is simply the period of time over which such a civilization might attempt to communicate with us once it has arisen.

Some of the factors in the equation are reasonably well known; such as the number of stars born each year in the galaxy, the percentage of these stars (like our Sun) that are hot enough, but also live long enough, to allow intelligent life to arise and the percentage of these that have solar systems. Others are far harder to estimate. For example, given a planet with a suitable environment, it seems likely that simple life will arise – it happened here virtually as soon as the Earth could sustain life. However, it then took several billion years for multicellular life to arise and finally evolve into an intelligent species. So it appears that a planet must retain an equable climate for a very long time.

The conditions that allow this to happen on a planet may not be commonplace. Our Earth has a large moon which stabilizes its rotation axis. Its surface is recycled through plate tectonics which release carbon dioxide, bound up into carbonates, back into the atmosphere. This recycling has helped keep the Earth warm enough for liquid water to remain on the surface and hence allow life to flourish. Jupiter's presence has reduced the number of comets hitting the Earth; such impacts have given the Earth much of its water but too high an impact rate might well impede the evolution of an intelligent species. It could well be, as some have written, that we live on a 'rare Earth'. How many might there be amongst the stars?

In addition, it has been widely assumed that once multicellular life formed, evolution would drive life towards intelligence, but this tenet has been challenged in recent years – a very well adapted, but not intelligent, species could perhaps remain dominant for considerable periods of time preventing the emergence of an intelligent species.

The final factor in this part of the equation is the percentage of those civilizations capable of communicating with us who would actually choose to do so. Our civilization could, but currently does not, attempt to communicate. Indeed there are some who think that it would be unwise to make others aware that here on Earth we have a nice piece of interstellar real estate! Any attempts at communication have to be made in the very long term – the round travel time for a two-way conversation would stretch into hundreds or thousands of years. It would be hard at present to obtain funding for such a programme. It is often cited that perhaps between 10% and 20% of civilizations would choose to communicate, but I suspect that this may well be highly optimistic.

The topic of 'leakage' radiation from, for example, radars and TV transmitters is often mentioned as a way of detecting advanced civilizations which do not choose to communicate. This, the author believes, unlikely. Any signals that could be unintentionally detected over interstellar distances are, by definition, wasteful of energy. Already, on Earth, high power TV transmitters are being replaced with low power digital transmissions, satellite transmissions are very low power and fibreoptic networks do not radiate at all. The 'leakage' phase is probably a very short time in the life of a civilization and one that we would be unlikely to catch. It could be that airport radars and even very high power radars for monitoring (their) 'near-Earth' asteroids might exist in the long-term, and give us some chance of detecting their presence, but we should not count on it.

When all these factors are evaluated and combined, the average time between the emergence of advanced civilizations in our Galaxy is derived. If we find it hard to estimate how often intelligent civilizations arise it is equally hard to estimate the length of time over which, on average, such civilizations might attempt to communicate with us. In principle, given a stable population and power from nuclear fusion, an advanced civilization could survive for a time measured in millions of years. Often a period of 1000 years for this 'communicating stage' is chosen for want of anything better. This length of time is critical in trying to estimate how many other civilizations might be currently present in our Galaxy. If, for example, a civilization arose once every 100 000 years – a reasonable estimate – but typically, civilizations only attempt to communicate for 1000 years, it is unlikely that more than one will be present at any given time. If, however, on average, they remain in a communicating phase for 1 million years then we might expect that nine other civilizations would be present in our Galaxy now.

When the Drake equation was first evaluated, the estimates of the numbers of other civilizations were quite high; numbers in the hundreds of thousands or even as high as 1 million were quoted. Nowadays astronomers who try to evaluate the Drake equation are far less optimistic. Many estimates are in the range of 10–10 000 but there are a minority of astronomers who suspect that, at this moment in time, we might well be the only advanced civilization in our Galaxy.

The truth is we just do not know. It was once said with great insight that 'the Drake equation is a wonderful way of encapsulating a lot of ignorance in a small space'. Absolutely true, but an obvious consequence is that we *cannot* say that we are alone in the galaxy. SETI (see below) is our only hope of finding out.

The author's own belief is that simple life will be widespread in the galaxy, but that few locations will keep stable temperatures for sufficient time to allow advanced civilizations to arise. Optimistically, their number might be in the tens to hundreds but it may very well be that none of these would choose to try to contact us so that we would remain in ignorance of their presence.

## 9.15.2    *The Search for Extra Terrestrial Intelligence (SETI)*

The SETI searches, observing in the radio part of the spectrum, have, as yet, only seriously searched a tiny region of our Galaxy. It will not be until the 2020s that the Square Kilometre Array, now on the drawing board, will give us the capability to detect radio signals of realistic power from across a large part of the galaxy. It is also possible that light, rather than radio, might be the communication carrier chosen by an alien race, but optical-SETI searches, seeking out pulsed laser signals, have only just begun.

In a 1959 paper in the prestigious journal *Nature*, Giuseppe Cocconi and Philip Morrison discussed the possibilities of detecting the existence of other civilizations by radio. Not only did they suggest some suitable nearby target stars, but also the optimum part of the radio spectrum in which to search for signals. They had no way of telling whether such a search would be fruitful but ended their paper with the following sentence: 'the probability of success (in our search for extraterrestrial life) is difficult to estimate, but if we never search, the chance of success is zero.'

## 9.16    The future of the universe

The accelerating expansion of the universe that is now accepted has a very interesting consequence. It used to be thought that with a slowing rate of expansion, as the universe became older we would see an increasing number of galaxies (as the distance we could see becomes greater). In a universe whose expansion is accelerating the exact opposite will be true – yes, we will be able to see farther out into space, but there will be increasingly less and less for us to see as the expansion carries galaxies beyond our horizon.

On the large scale, the space between the galaxy clusters will be expanding – carrying them ever faster apart – but it is believed that clusters like our own Local Group will remain gravitationally bound and, in fact, its members will merge into one single galaxy largely made up from our own Milky Way Galaxy and the Andromeda Galaxy. If one looks forwards in time to ~100 billion years, any observers in existence within this 'galaxy' would see a totally empty universe! The expansion of space will have carried all other galaxies beyond our horizon – the edge of the visible universe.

It would be virtually impossible for such observers to learn about the evolution of the universe for a number of reasons including the fact that the peak of the energy spectrum of the CMB will have red shifted down to ~1 m. A uniform radio background at this wavelength could, in principle, be detected, but its intensity will have reduced by 12 orders of magnitude so that it would be virtually impossible to detect!

From theoretical studies of stellar evolution and how the relative abundances of the elements change with time (for example, the amount of hydrogen is reducing and that of helium increasing as a result of nucleosynthesis in stars) it might well be possible to estimate the age of the galaxy, but it would be not be possible to infer that its origin involved a Big Bang. It would be tough being an astronomer!

We happen to live at the only time in the history of the universe when the magnitude of dark energy and dark matter are comparable and also when the CMB is easily observable; this enables us to infer the existence of dark energy, the way in which the universe has evolved since the Big Bang and its future in a runaway expansion (Figure 9.15).

Any observers present when the universe was young would not have been able to infer the presence of dark energy as, at that time, it would have had virtually no effect on the expansion rate. Those in the far future will not be able to tell that they live in an expanding universe at all, and not be able to infer the existence of dark energy either! As the longest lived stars come to the end of their lives, the evidence that lies at the heart of out current understanding of the origin and evolution of the universe will have disappeared.

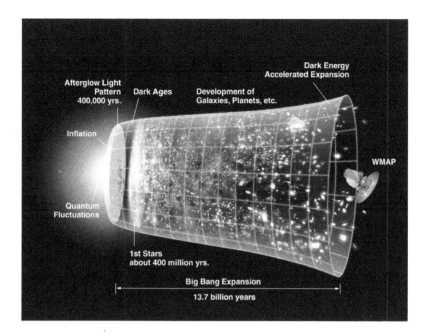

**Figure 9.15**  A time-line of the universe. Image: NASA/WMAP Science Team.

It is interesting to note that we live in what is perhaps the very best time in which to explore the mysteries of the universe. As Lawrence M. Krauss and Robert J. Scherrer have stated: 'We live in a very special time in the evolution of the Universe: the time when we can observationally verify that we live in a very special time in the evolution of the Universe!'

This is a wonderful time at which to study astronomy.

I trust that this book might have helped you on your way.

# Index

References to *figures* are given in *italic* type; references to **tables** are given in **bold** type.

CPSIA information can be obtained
at www.ICGtesting.com
Printed in the USA
BVHW050124251118
533782BV00003B/5/P

9 780470 033340